Single Lens

BRIAN J. FORD

Single Lens

THE STORY OF THE SIMPLE MICROSCOPE

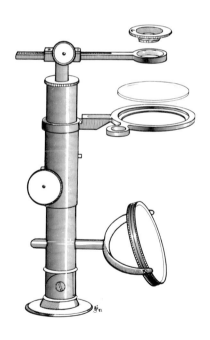

HEINEMANN: LONDON

William Heinemann Ltd
10 Upper Grosvenor Street, London W1X 9PA
LONDON   MELBOURNE   TORONTO
JOHANNESBURG   AUCKLAND

First published in Great Britain 1985
© Brian J. Ford 1985
SBN 434 26844 5

Photoset by Rowland Phototypesetting Ltd
Bury St Edmunds, Suffolk
Printed in Great Britain
by Billing & Sons Ltd, Worcester

Contents

Illustrations

Preface and Acknowledgements

Early in 1981 at the Royal Society I came across nine specimen packets dating back three centuries, to the earliest days of microbiology; several weeks later I assembled at my laboratory a group of microscopes from the early nineteenth century to calibrate and compare them. Investigations showed that the specimens, and the instruments, were of a far higher quality than orthodox opinions lead us to suppose. The first results of the work were published by the Royal Society in July 1981, and the microscopes were first demonstrated at an Extraordinary General Meeting arranged by the Linnean Society in November of that year. During September 1982 a Special Public Lecture was arranged by the British Museum of Natural History in South Kensington and this book outlines the work carried out up to that time, setting it into its historical context. However, before the story of some of the great pioneers of the microscope is told, I wish to acknowledge the large amount of assistance which has been offered by friends and colleagues in several countries. Among those who offered help and the hospitality of the facilities in their charge were: the late Peter

Hans Kylstra, Director, University Museum of Utrecht, and J. van Zuylen of Zeist, Netherlands; G. Van Steenbergen and the staff of the Henri Van Heurck Museum of Historic Microscopes, Royal Zoological Society, Antwerp, and J. Hansen – Van Walle of Antwerp, Belgium; P. S. Green and Gren Ll. Lucas of the Herbarium and David Cutler of the Jodrell Laboratory, Royal Botanic Gardens, Richmond, Surrey; William T. Stearn, then President, Gavin Bridson, Librarian, Elizabeth Young, Secretary and Gina Douglas (now Librarian) of the Linnean Society of London; Horace Dall of Luton, Bedfordshire; the late C. R. Burch, F.R.S. of Bristol; Gunnar Broberg of the Department for the History of Science and Ideas, University of Uppsala, Sweden; N. H. Robinson, Librarian, and L. Townsend, formerly Archivist, of the Royal Society, London. Additional help during my visits has been given by the staff of the Boerhaave Museum, Leiden, the Conservatoire des Arts et Métiers, Paris, the Technical University of Delft, the Deutsches Museum of Munich, the Museum of the History of Science and the Bodleian Library of the University of Oxford, and the Victoria and Albert Museum, the Institute of Biology, the Science Museum, the Natural History Museum, the Quekett Microscopical Club, the Royal Microscopical Society and the British Museum, during my work with them. For advice on the Dutch language and for assistance with translations I wish to thank Beverly Collins of Zeist and J. A. van Dorsten, Professor of English Literature at the University of Leiden. During the Utrecht experiments I was helped by Robert Frederik with Jaap Stolp, Pieter Smiesing and Karel Snethlage and the technical staff of the University Museum; and at the Royal Society I was grateful to have the advice of Sir Andrew Huxley, President; Sir William Paton F.R.S., R. V. Jones F.R.S. of Aberdeen University and John Phillips F.R.S. of the Zoological Society of London. As Visiting Scientist at University College Cardiff, I have been encouraged by Denis Bellamy, Professor of Zoology, and I am grateful to Carole Winters and John Morgan for technical assistance. D. A. Griffiths of the Ministry of Agriculture, Fisheries and Food offered advice on the identity of the mites, and Adrian Amsden of the National Museum of Wales assisted with entomology. The late Derek de Solla Price of Yale University advised on his earlier attempts to resolve the fate of the collection of microscopes missing since *c*.1820 from the Royal Society, and I was helped by advice from the late George W. White of Richmond, Surrey. W. A. S. Burnett of Dorking, Surrey, and Irene Manton F.R.S. of Leeds University advised on the condition of Robert Brown's microscope at Burlington House and I received further help from Stanley Travers, Industrial Photographer, of Cardiff, John S. Slade, senior microbiologist, Thames Water Authority, and John Hopkins, Librarian, Society of Antiquaries. Savile Bradbury of the University of Oxford was kind enough to read the manuscript, and I received further invaluable advice on the Leeuwenhoek microscopes from J. van Zuylen, and on the botanical microscopy of the nineteenth century from William T. Stearn, both of whom spent much time working through my notes prior to publication. Grants in aid of research were kindly made available through the Royal Society, the Appleyard Trust of the Linnean Society of London, the Kodak Bursary Scheme, the Spencer-Tolles Fund of the American Microscopical Society, and the Botanical Research Fund. To all these, and the many other friends who offered counsel, advice and encouragement during this exciting but complex programme of research, my heartfelt thanks.

Brian J. Ford
London, 1984

Introduction

I am cradling in curved fingers a small piece of metal, brown, mellowed with the passing of time, and utterly unprepossessing. It would be difficult to imagine that any man-made object could be more ordinary. Only a very small proportion of people would be able to guess what it was, and those who did know the answer (scientists, in most cases) would have little detailed knowledge about it.

Yet this postage-stamp-sized object is one of the most important instruments of scientific discovery. It changed our lives, altered our self-image, revolutionised our understanding of the world in which we live. Our modern biology-oriented era of the cell nucleus, bacteria and the whole gamut of microscopic organisms arose through the use of this little object. For it is a microscope – not the first microscope, nor the most powerful; but an instrument with a significance that few people have ever begun to appreciate.

It is easy to imagine your incredulous reaction: 'A microscope? This little scrap of metal? But surely, a microscope is a large and heavy metal object with milled knobs, wheels, gears, racks, pinions and stainless steel – and with complex arrangements of lenses that take an invisible object and reproduce it as a room-sized image through the might of modern technology . . .' Well, yes, that is how we do think of microscopes today. Like motor-cars or video sets, a modern microscope has to look impressive if it is to perform its task properly.

Today's typical laboratory microscope is so ingrained in our minds from those television images in science documentaries that it is easy to see why people cannot imagine an optical microscope in any other form. The little die-cast and plastic toy microscopes that children have as birthday presents look much the same shape as the full-sized laboratory instrument. There are obvious practical reasons for that.

But in this book I want to show you a different kind of microscope altogether. Instead of complicated arrays of lenses grouped together in precision mountings, I present just one, single lens. Nothing more than a minute magnifying glass, typically not much larger than the head of a pin, yet capable of producing an image that is only a little lower down the scale of magnifying power than a modern instrument. Microscopes with many lenses are known as compound microscopes, and the instrument through which we are going to look in this book is the largely forgotten type with just a single lens. Single-lensed microscopes is one term for them, but the more correct expression is *simple* microscopes, in distinction from the *compound* instruments with which we are familiar today.

Most of the people who used simple microscopes knew little about the microscopic universe when they began their work, for the obvious reason that nobody else knew very much, either. Between them they discovered most of the fundamental microscopic phenomena we take for granted. The popular idea many people have – namely, that it was the upsurge in lens technology of the Victorian era that gave us today's insight into microscopic phenomena – can be dismissed.

For all their simplicity, these single-lensed microscopes could demonstrate many of the objects we associate with more recent instruments. Indeed, with the tiny brass microscope I am holding at this moment, you can see cells, nuclei, bacteria . . . the whole microbe world can be glimpsed with nothing more complicated than a speck of glass set in a sliver of metal.

None of this was at all surprising for the pioneers, who knew full well what these diminutive magnifiers could achieve. And today's specialist is well aware of the theory behind the simple microscope. But how was that theory translated into practice? That is what still puzzles some people. Time after time, in the literature that covers the development of modern microbiology, you read of authors who add how 'surprising it is that so much seems to have been seen', or who describe discoveries in detail before adding that the work was done with 'nothing more than a simple microscope' – concluding that how much was really seen, and how much more was intelligent guesswork, remains a mystery.

So let me come clean at once about the little rectangular microscope I have already mentioned. It was made around 1700 in Delft, Netherlands, by a draper named Leeuwenhoek. His name, as it has come down to us, is Antony van Leeuwenhoek; but he was christened 'Thonis', and acquired the 'van' as an affectation in 1685 when he was fifty-two. He had a normal secondary education, was apprenticed to a draper in Amsterdam, and lived for his adult years in the pleasant and quiet city of Delft, much of which has changed relatively little since Leeuwenhoek's time.

On his own, and in his spare time, he founded the science of microbiology. It was Leeuwenhoek who drew cells and, with them, the cell nucleus; it was he who documented spermatozoa; it was Leeuwenhoek who discovered the microbe world and first observed bacteria. The kind of microscope he used was hand-made, sometimes being fashioned from metals he refined himself and then beat into shape. The lenses he made himself from fragments of glass. The microscope I am considering here is actually in Utrecht, at the University Museum. It has been brought from a locked safe (for the 'Leeuwenhoek microscope' in the display cabinet is only a replica: the original is too precious by far to leave lying in a glass case) and carefully tipped from its little cardboard box.

There is a single lens set into the metal plates of the microscope body. This bead of glass has a magnification of 266 times, at which rate a bluebottle fly would be a metre in length and even a smaller-than-average bacterium would be clearly visible as a round dot the size of this full stop. The moving bacteria of ponds, and the swimming organisms that live between our teeth, all show up with clarity.

Leeuwenhoek used a series of simple screws to move the specimen across the field of view, and to focus its image; and

FIG. 1   The Leeuwenhoek microscope at the Royal Zoological Society of Antwerp. The single lens is housed inside a prominence beaten into the brass plate, whilst the specimen was fixed to the pointer in front of the lens. The screws were used for positioning and focusing the specimen. The history of this particular example is discussed on p. 64.

yet this simple little instrument is still capable today of magnifying as well as you would need for most modern scientific purposes. The highest magnification that can be obtained with a modern microscope is around a thousand times – and here we have a little hand-made microscope some three centuries old that can itself reach to one-quarter of that.

This itself is a fact worth pondering. These days, when our laboratories are jam-packed with ultra-sophisticated equipment and automated microscopes are as commonplace as washing machines, we take high magnification for granted. How salutary it is to realise that Leeuwenhoek's own amateur version could give results that rate so highly that the best modern equipment is only four times more powerful. The performance of modern cameras, modern aircraft, modern telescopes, modern ships, can often be hundreds of times better than their earlier counterparts. Yet today's optical microscope is only four times better than Leeuwenhoek's primitive version. What an achievement that was!

Remember that Leeuwenhoek's microscopes were not the first. That is an important point to emphasise at the outset. Many descriptions of microscopes show the simple type of instrument he made, as it were, at the base of the microscope's family tree, so that the complicated constructions of more recent years spring from this simple beginning. But that is an error.

Microscopes were in existence for perhaps half a century before Leeuwenhoek made his first example. What is more, the recognisable modern shape of microscope, with a body tube, an objective lens near the object and an eyepiece for viewing, was already in existence before Leeuwenhoek went to school. The point of interest is not that his microscope was earlier, just that it was *better*.

A second and perhaps more common misrepresentation is that the simple microscope was a temporary aberration, a short phase in the history of microscopy that is scarcely worth considering at length. That too is wrong. Simple microscopes were still in popular use up to the middle years of the last century, so the instrument had a history that lasted for at least 200 years. As late as 1854 the Society of Arts awarded a prize for the design of a new simple microscope, as we shall see, and a Zeiss doubtlet lens of the same era was found to be 'significantly inferior' to the lens in the Leeuwenhoek microscope.

Why then have we heard so little of simple microscopes? I think a large part of the answer lies in the fashions that come and go in science, and also in the snobbishness of science. Simple microscopes always had a bad press. They were 'crude' or 'plain', 'ordinary' or 'basic'. The Victorian compound microscope, by contrast, was complicated and glamorous and just the instrument to boast about.

In consequence, even the most beautifully designed and aesthetically pleasing of the simple microscopes have been overlooked by history and by science. In this book we will follow the development of this revolutionary little instrument until it was in demand by such great biologists as Charles Darwin, was commissioned by the Royal Family, and laid the foundations of our modern era of biology through the work of such people as Bentham and Hooker, and Robert Brown (from whom Brownian movement obtained its name). One micro-scope-maker produced instruments for all these people, and for many more; yet he has been frequently overlooked by histo-rians. Whilst his followers, who made fine-looking microscopes with celebrated names – Ross, Powell & Lealand and the rest – are known to many collectors of antiques and feature in all the standard reference works on microscopes and their history, this man – Robert Bancks – is unfamiliar; usually just a surname, if that.

Yet with his microscopes, far more fundamental discover-ies were made. Only once have I found a detailed published account of one of his microscopes, and even that attributes it to another maker.

The pioneers in more fashionable arenas are well documented and also well understood. I have studied the first-ever photographs taken with microscopes; handled with infinite care the oldest bound books; marvelled at pioneering television sets; peered at culture plates where antibiotics were first discovered. I have turned the pages of well-known and historic writers, and been regaled with embarrassed frankness by the man who saw the keeper of the original thermometer of Celsius take it from its case for an exhibition in Uppsala, lay it on a chair, and then in an unguarded moment sit on it, breaking this irreplaceable relic into three pieces. Artefacts from other areas of interest are much better known.

Yet the simple microscope remains beyond the pale. Twenty-six of Leeuwenhoek's microscopes are very likely still in existence, lying in a dusty attic in London, unrecognised. Of the nine known to be associated with his name, one was discovered a few years ago in a discarded box of laboratory oddments. Other microscopes of the greatest historic import-ance have turned up in cupboards, in private collections, and one was returned to a learned society in London after being bought up in a house sale in Dorset.

There is every reason to rehabilitate the simple micro-scopes. As historic relics they have a fascination of their own, but as an overlooked facet of scientific history their interest is paramount. In fact there is a further theme of the discussion – there is good reason to believe that this much-misunderstood, widely ignored instrument has a more important niche in our history than even these considerations might suggest. If we realign the way the construction of the microscope is seen to have developed, then the simple microscope emerges as the clear ancestor to today's laboratory instrument. The principles of design were not merely peculiar to simple microscopes, but laid the foundations for the magnificent-looking brass and lacquer constructions of a hundred years ago. More important still, the designs of today's research microscopes have more in common with the simple microscopes of the early 1800s than with the compound microscopes of the early days.

No single book like this can encompass all the varied shapes and sizes through which the simple microscope developed and grew over the years. It would be a pleasure to feel that here was a definitive list of all the attainments of each pioneer, and all the instruments they used; but the listing of terse and clinical data like that is a poor read. One of the first of the historians of early microscopes, Henry Baker, wrote in his *Microscope Made Easy* of 1743: 'There are many pretty little Contrivances sold at the Shops for the viewing of small Objects, which are entertaining as far as can reasonably be expected from them: but to enumerate all these would be a tedious task.' Those words (though with 'task' sometimes mistakenly rendered as 'talk', because the old-fashioned form of the letter 's' looked in that context more like 'l') were used a decade ago by a President of the Royal Microscopical Society and a Professor at Galveston, Texas, John Bunyan, introducing an earlier publication of mine on the development of microscopy, and they embody a useful hint to anyone following in Baker's footsteps.

Rather than a catalogue, a kind of historical shopping-list, let us take a broad sweep through the development of this Cinderella of scientific instruments and relate it to its context in time. I will take the opportunity to relate something of the people who used it, and we will see first hand the work they did. Starting with the greatest of the first pioneers, Leeuwenhoek, and working through to the last of the leading microscopists to base his work on the single lens – Robert Brown, who died in 1858 – we will see how important was the work performed with the simple microscope.

In the end, you will, I hope, be left with a taste of the way this unsung hero of scientific instrumentation matured and developed. And you will have glimpsed a little of the background to one of the most important and far-reaching areas of scientific endeavour – one which has been, in the past, all too often forgotten.

1 | Origins

The fact that light always travels in straight lines was taught to us as children. It was a principle known well enough to the ancients, who probably taught it to their offspring, too. Euclid – who wrote about 300 BC – imagined that vision was due to rays shining out of the eyes, rather than into them; but the first of his *Suppositiones* read:

> [*Visual*] *rays issue from the eye and proceed in straight lines a certain distance apart.*

Here, from two and a quarter millennia ago, is the 'straight line' doctrine expounded. Common experience tends to make us question the principle. Everyone has seen a straight stick pushed into water at an angle, looking bent at the water surface. Whether it is a cocktail stick holding a delicately poised olive, or a pole poked into a sea-side rock-pool, the phenomenon of bending is well known to all of us.

So – though light rays may travel in lines as straight as a die – it is equally clear that they can bend. Though everyone knows that they do, it may be you are not so sure *why* they do it. The effect is known as refraction, and it takes place whenever a beam of radiation shines at an angle into a medium where it travels at a different velocity.

A familiar example would be a line of walkers, passing across a meadow in a straight row abreast, each holding on to the person on either side. Here is your beam of light moving in a straight line. But what happens when this line of people encounters a ploughed field? The first people to step into the rough ground will begin to move at a lower speed. When their neighbours reach the rough ground, so do they. As the straight line of people strikes out across the more difficult terrain, the drop in speed will have brought about an equivalent change in direction. This is refraction. The velocity of light in glass is less than when light travels through air, and so the wave-fronts in the light beam are swung round just like the row of marchers in our anthropomorphic analogy. The effect on a beam of light entering glass at an angle is to make it change direction. If the glass surface is curved, then the further along the surface the light rays pass through the interface, the greater will be the angle through which they are bent.

That, in a few words, is how a lens works. We can go back to the writings of Claudius Ptolemy in the second century BC to find the principles of refraction expounded. He wrote:

> . . . the flexure is of the same magnitude in each of the two kinds of passage, but differs in species; for, whereas in the passage from the rarer to the denser the flexure is towards the perpendicular, in the passage from the denser to the rarer the flexure is away from the perpendicular.

He proved his scientific abilities by building a brass measuring circle to enable him to find experimentally how a beam of light changed direction when it was shone into water. If the beam was at 10° to the water surface, he found the angle of refraction to be 8° (modern calculations would make that 7½°); if the angle of incidence was 20°, Ptolemy's measured angle of refraction was 15½° (modern figure 15°) and so on. Almost all his measurements were within half a degree of the figure we would cite today, a considerable attainment for someone working 150 years before the birth of Christ.

Refraction of a beam of light is only the first stage towards an understanding of magnification. Long before the idea of a lens began to emerge, the ancients were familiar with the focusing effects of a concave mirror. It was said that Archimedes of Syracuse used focusing mirrors to direct an image of the sun on the ships of the Roman fleet when they sailed into the harbour at Syracuse in 214 BC. Whether this story is true or not remains a matter of scholarly debate. Georges Louis le Clerk, Count de Buffon, a somewhat cranky experimenter of the eighteenth century, claimed to have constructed a rig of 128 mirrors which was capable of igniting wood at a distance of 50 metres, which has been taken as supporting the legend of Archimedes. However, against that you may cite the philosopher-writers Livy, Plutarch and Ploybius – historians who were all near-contemporaries of the episode – who made no reference to the great feat. The legend of the mirrors was first written up by the designer of the Saint Sophia cathedral in Istanbul, Anthemius, in the sixth century AD, eight hundred years or so after the event would have taken place. It is not a record that gives confidence in its historical veracity.

After the time of Ptolemy, the development of optics seems to have halted altogether. Brief forays into the story-telling of legends by Anthemius hardly count, and it was not until the upsurge in Arabian science in the ninth century AD that the next steps were taken. At the town of Basra, in what is now Iraq, some 50 miles from the Persian Gulf, about AD 695, a notable philosopher–scientist named Al-hazen was born. By the time he died in Cairo in the year 1039 he had carried out countless experiments in optics and compiled a work that became known as the *Optical Thesaurus* (originally *Opticae Thesaurus Alhazeni Arabis* in the familiar Latin translations). Though he owed much to the works of his predecessors – for the experiments of Ptolemy and Euclid were well understood in Persia at that time – he showed a reliance on experimental proof that did much to stimulate the movement towards our own scientific era.

Altogether, the *Optical Thesaurus* is divided into seven books, containing around 450 propositions. Al-hazen began straight away by breaking new territory, describing in detail the anatomical structure of the human eye. He recognised the three humours of the eye that entered Western anatomical teaching – the aqueous, vitreous and crystalline humours – and the four layers of the eye now familiar to anatomists. He understood the difficulty of having two eyes seeing one image, and assumed that the place where the two optic nerves meet in front of the brain was the site at which the images merged:

> the two images are received by the two eyes and combine at the meeting-point of the two nerves – and thus the image of a single object will result.

But it was in Book VII that Al-hazen dealt with refraction at some length. Here the idea of a lens began to make its appearance – not in a form that made optical instruments predictable, quite, but at least in laying the groundwork. His words now seem to be so close to the concept of a magnifying lens that they count, at least, as a near miss. In Proposition 44 he writes:

> if the object be symmetrically placed in the circle between the centre and the eye-point, then the image will seem greater than the object.

For some inexplicable reason, Al-hazen's reasoning took him only to a consideration of refraction at *one* surface so his idea, though it explained magnification of an object inside a wine-glass of water, said nothing on the action of true lenses.

Yet how near Al-hazen came! Look at these words from Proposition 49 of Book VII.

> If the eye-point, the centre of a refractive convex sphere denser, and the visible object beyond the sphere be in the same straight line, the image . . . *will be larger than the object.* (my italics)

Here are the roots of magnification theory, without the practice.

And what of the artefacts that persist from earlier ages? There are many lens-like polished semi-precious jewels from ancient times which occasional writers have claimed to be examples of early lens-making. In each case that they have been examined closely they turn out to be nothing more than pendants or baubles that were shaped as they are for practical reasons, and not because of a desire to make them useful as optical elements.

Perhaps the best known, for it has been written up by Erich von Däniken as a remnant of technology brought to us from space-travelling visitors, is the Assyrian lens in the collections of the British Museum. This object is a biconvex structure polished from crystal. To read the words of von Däniken it is a super-miraculous artefact betokening cosmic influences over our predecessors: 'To grind such a lens requires a highly sophisticated mathematical formula,' he writes. 'Where did the Assyrians get such knowledge?' It might seem churlish to say at the outset that grinding a lens does not require a sophisticated mathematical formula. Countless lens-makers of the past worked by trial and error and produced very good results. A normal lens has its surface shaped like part of a sphere which, even if difficult to attain if you are an amateur at the job, is not a 'sophisticated formula'. And the pioneers often used surfaces that were not perfectly spherical, as a result of their primitive techniques. These aspheric lenses gave results that were frequently rather better than would have been the case if they had been made as perfectly spherical magnifiers. So even when the true sphere was unattainable, the results were liable to be improved, and not degraded.

However, one does not take issue with the von Däniken hyperbole because of that kind of misstatement. The fact is that the so-called lens is nothing of the sort. It was examined in detail in 1883 (when an account was published in the *Journal of the Royal Microscopical Society*) and again by John Mayall junior in 1885, who argued strongly against its ever being usable as a lens. There was a later account by W. H. Dallinger which was clear in its dismissal of the idea:

> The remarkable piece of rock crystal . . . is oval in shape and ground to a plano-convex form, and was found by Mr Lanyard during the excavations of Sargon's Palace at Nimroud; Sir David Brewster believed it was a lens designed for the purpose of magnifying. If this could be established it would of course be of great interest, for it has been found possible to fix the date of its production with great probability as not later than 721–705 B.C.
>
> There are cloudy striae in it, which would prove fatal

for optical purposes, but would be even sought for if it had been intended as a decorative boss; while the grinding of the 'convex' surface is not smooth, but produced by a large number of irregular facets, making the curvature quite unfit for optical purposes.

There are other old lens-like artefacts. In the Egyptian Gallery of the British Museum are three glass objects that look like lenses, and at University College London there was reputed to be another pair of similar objects. They were carefully scrutinised by Sir W. M. Flinders Petrie in the 1880s, who showed that they were of little value as magnifiers, and the conclusion must be that they are nothing more elaborate in purpose than polished tablets that have become separated from their original jewellery mounts.

Around AD 1200 there was a sudden resurgence in learning, and it was then that the subject of optics began to re-emerge. By 1200 there were universities at Oxford and Cambridge, at Paris, at Bologna and Salerno. One of the new generation of thinkers who left some tantalising writings on the subject was Roger Bacon. He joined the Franciscan Order as a young man in his thirties. The climate of the time did not encourage random scientific experimentation, so he concentrated instead on private philosophy and on teaching students.

The need to keep out of trouble meant that much of Roger Bacon's writing was in a contorted kind of philosophical code, in which ideas were hinted at, rather than expressed. Interwoven with his own ideas were references to the received wisdom of his age, replete with metaphysical references to dragons, miracles and supernatural longevity. The fact that Roger Bacon describes the best way to entice a dragon from its cave and saddle and harness it for riding is enough to make anyone turn from his writings with more than a trace of suspicion!

But what of these words? In his great work *Perspectiva*, written around 1267 (and part of an opus requested by the then Pope, Clement IV), he explains:

[Great] things can be performed by refracted vision. The greatest things may appear exceeding small, and on the contrary; also the most remote objects may appear to be just at hand, and on the contrary ... Thus from an unbelievable distance we may read the smallest letters, and may number the smallest particles of dust and sand by reason of the greatness of the angle under which we see them; and on the contrary we may not be able to see the greatest bodies just near us by reason of the smallness of the angles under which they may appear ... Thus also the sun, moon, and stars may be made to descend hither in appearance, and to appear over the heads of our enemies: and many things of the like sort which would astonish unskilful persons.

If the letters of a book, or any minute object, be viewed through a lesser segment of a sphere of glass or crystal, whose plane base is laid upon them, they will appear far better and larger.

His descriptions continue. You might at once conclude that Bacon knew all about telescopes and magnifying lenses, and a good number of historians have reached that conclusion. Yet a closer examination does not substantiate that interpretation. To begin with, all of Bacon's diagrams show that he is talking – like Al-hazen – of objects that are *inside* magnifiers, like a stick in a goblet of water. Letters, for instance, are not enlarged by a double-convex lens held above them, but by the 'segment of a sphere' that is 'laid upon them'. The glass dome is simply laid onto the page, and the letters looked at from above.

The descriptions of the magnifiers that look rather like telescopes, from the evidence of the printed page, were probably based on work with concave mirrors rather than lenses. In one telling phrase Roger Bacon describes how the magnified image can make 'a small army appear a very great one'. No telescope – no matter how effectively it may magnify the dimensions of an object – can increase the numbers of individuals you see! A series of reflections from a system of mirrors, on the other hand, can do that perfectly well, as we all know from experiments with Mother's dressing-table mirrors as children, which could generate an endless cascade of visions of ourselves, streaming away towards infinity.

Roger Bacon did not describe a compound telescope, as far as we can tell. But he did leave us, from the middle years of the thirteenth century, an unmistakeable description of the use of a spherical magnifier as an aid to reading. In that sense he did record the idea of a single, simple lens. It was a lens with only one curved side, not a double convex magnifier; but the principle had in those words entered both the language and the intellectual repertoire of science.

All the while, there was the faintest undercurrent of interest in lenses that were not magnifiers – i.e. convex structures – but *concave* lenses. If you are short-sighted, your eye acts as though its lens were too powerful. To correct this trend, you require a *negative* lens – and negative lenses are concave. There is one tantalising reference from Pliny, when he describes what may be the earliest use of a negative lens as an aid to vision:

> Emeralds are usually concave so that they may concentrate the visual rays . . . when held supine they give images of objects the same way mirrors do. The Emperor Nero used to watch in an emerald the gladiatorial combats.

Now there *is* a teaser! The reference is too ambiguous for us to know what Pliny meant: was the Emperor looking at a reflected image in the surface of the emerald? Was he merely finding the green colour kind on his ageing eyes? Indeed, was it an emerald he was using at all (one commentator has argued that the emerald was unknown to the Romans at the time)? To my mind the possibility that Nero was using a small plate of green glass, and not a gem-stone, is neither here nor there. Pliny's words contain enough to make one reasonably confident that the Emperor was gazing through a concave lens, which would have given him a crystal-clear view if he was becoming myopic in his old age. Any short-sighted person will know from experience that a concave lens has a miraculous effect on the clarity of vision, and it may be that Nero was the first-recorded person in history to use a monocle.

By the latter part of the thirteenth century, spectacles were in existence. It has been claimed that they were invented in Florence around 1280. A monk in Pisa wrote in 1305 that he had personally known the person who invented spectacles about twenty years earlier (i.e. 1285) and a Florentine nobleman named Amati had engraved on his tombstone in 1317 that he had been the inventor of spectacles many years before, but had kept the fact a secret from all but a few friends.

Franciscus Maurolycus (1494–1577) published the first work which attempted to systematise the theory of lenses in spectacles. In his work entitled *Photismi*, and published in 1521, he described light in terms of straight rays of differing intensity – a clear step away from the traditional view, namely that sight was due to radiation emitted by the eye – and in his *Diaphana* of 1553 he coined the idea that the lens in the eye was analogous to a glass lens. His conclusions over the correction of faulty eyesight by the use of glasses have a decidedly scientific style. He reasons that in a long-sighted person the lens of the eye has too small a curvature (*expansior pupillae facies*), and therefore a convex lens is needed to correct the fault. The short-sighted person, Maurolycus concludes, has a lens that is too highly curved, and hence refracts the light so much that a negative or concave lens is necessary for correction. This view is not far from the modern understanding.

He also realised that earlier writers – including Bacon – must have been wrong in asserting that the rays of light that entered the eye were those perpendicular to the pupil. Maurolycus states that this is absurd, and points out that all the rays reaching the eye from the object (apart from the ray on the axis – that is, the ray that runs in a straight line from the object through the lens) must be refracted.

Here the groundwork was being done for the later development of optical instruments. No one can ever know who was the first person to construct a magnifier *per se*. The earliest accepted description of an optical instrument in use lies in the 1571 edition of *Pantometria*, published in London by Henry Bynneman in the year of the author's death. This work was written by Leonard Digges, and was completed and prepared for printing by his son Thomas. In it appear the following words:

> Marvellous are the conclusions that may be performed by glasses concave and convex of circular and parabolicall formes. By these kind of glasses, or rather frames of them, placed in due angles, yee may not onely set out the proportion of an whole region, yea represent before your eye the lively image of every Towne, Village, &c. and that

in as little or great space or place as ye will prescribe, but also augment and dilate any parcell thereof, so that whereas at the first appearance an whole Towne shall present itself so small and compact together that yee shall not discerne any difference of Streates, ye may by application of Glasses in due proportion cause any peculiare House, or Roume thereof dilate and shew itself in as ample forme as the whole Towne first appeared, so that ye shall discerne any trifle or reade any letter lying there open, especially if the sunne beames may come unto it, as plainely as if you were corporally present, although it be distante from you as farre as the eye can decrie.

That is a clear description of a telescope in action. I can offer an earlier candidate than this, from a man who coined the notion of the germ theory of disease three centuries before Louis Pasteur, and who knew about compound lens systems long before Maurolycus was writing about spectacles. This was Fracastoro of Verona, who included the following passages in his *Homocentrica* of 1535:

> If anyone should look through two spectacle glasses, one being superimposed on the other, he will see everything much larger and nearer . . . Certain lenses are made of so great a density that if anyone should look through them either at the moon or at any one of the stars, he would judge them to be so near that they all but surpass actual towers.

This intriguing little passage is, in my view, of the greatest importance. The words 'two spectacle glasses . . . superimposed' is a clear indication of a compound lens system, and the fact that you might look at the moon or stars through lenses and make them 'to be so near' is an unmistakable reference to a telescope of some kind. So the idea of a compound optical magnifier was current in 1535. That is four and one half centuries ago, almost half a millennium!

During the sixteenth century it was known that when you look through a positive lens held not too far from the eye, the world seems enlarged and nearer but indistinct, whereas when you look through a negative lens in the same position the image is diminished but very clear. There was some speculation about how to combine the phenomena, but understanding at the time was insufficient to provide an answer. It was possible to combine a powerful positive lens with a weak negative lens, but the real breakthrough came when it was realised that the answer lay in using a combination of a *weak* positive lens and a *strong* negative one – coupled with the realisation that the distance between the lenses was also important. This development came about in the late sixteenth century, most probably in Italy.

By 1590 compound microscopes probably existed. The inventor has traditionally been claimed as Zacharias Janssen of Middelburg, though at the frequently quoted date of 1590 he would have been an infant. An old instrument that was associated with Janssen was examined by Pieter Harting, a professor of the University of Utrecht, in 1867. He found that it was indeed a primitive kind of compound microscope and magnified up to nine times.

One of Zacharias Janssen's childhood friends eventually became the Dutch Ambassador to the court of Louis XIV, and in a letter dated July 1655 he described one of those microscopes:

> It was not (as such things are now made) with a short tube, but one almost a foot and a half long, the tube itself being of gilded brass and of two inches diameter, mounted on three brazen dolphins likewise supported on a basal disc of ebony.

The compound microscope gradually acquired fashionability. In 1625 Francesco Stelluti published a series of views of the honey-bee, showing the details of antennae, limbs and proboscis revealed through a compound instrument (see p. 20).

Then, in 1665, one of the most influential books in the history of the microscope was published in London. It was entitled *Micrographia*, and it stimulated for the first time a wave of public interest in the microscopic world. The author was Robert Hooke, commemorated for most of us through 'Hooke's Law', which relates that the expansion of a spring varies with the pull applied to it, and one of the least profound observations he ever made. In *Micrographia* he clearly intended to write

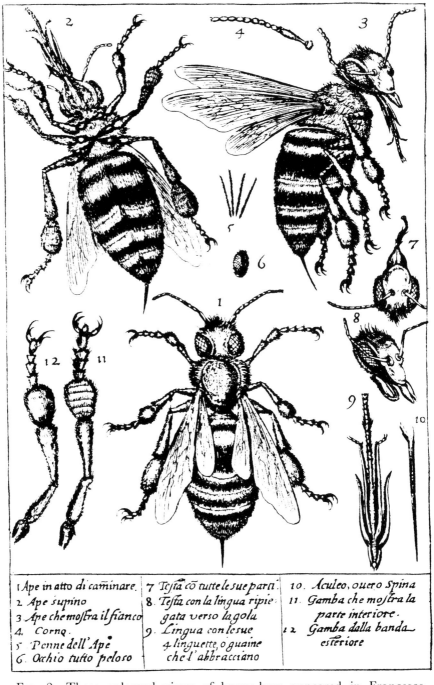

1 *Ape in atto di caminare.*	7 *Teſta cõ tutte le sue parti.*	10. *Aculeo, ouero Spina*
2 *Ape supino*	8. *Teſta con la lingua ripie-*	11. *Gamba che moſtra la*
3 *Ape che moſtra il fianco*	*gata verso la gola.*	*parte interiore.*
4. *Corno.*	9. *Lingua con le sue*	12 *Gamba dalla banda*
5 *Penne dell' Ape*	*4 linguette, o guaine*	*eſteriore*
6. *Ochio tutto peloso*	*che l' abbracciano*	

FIG. 2 These enlarged views of honey-bees appeared in Francesco Stelluti's 'Descrizzione dell' Ape', and are the earliest known studies made with the aid of a microscope. They were first published in 1625.

a bestseller. The book's preface states that he began by ensuring that nobody else was likely to write a similar volume before he started work. Christopher Wren, whom we think of as an architect but who was also a brilliant practical scientist, was one of those possible authors, and Hooke made sure to check with him first that there was nothing similar in the pipeline for publication.

Robert Hooke was a remarkable man, inventive, incisive, with a rare quality of perception and a brilliantly broad mind. He was plagued with ill-health and was, by all accounts, physically unattractive, with long untidy hair, a stooping gait and grey eyes, his complexion pallid and his walk rapid, yet he had a quality of intent and concentrated mental agility. He was born in Freshwater on the Isle of Wight in 1635 and, after gaining an M.A. at Oxford, he was appointed assistant to Robert Boyle. One of Hooke's experiments, on capillary attraction, came to the attention of the newly formed Royal Society in 1662 and in November of that year he left Boyle's service to become the Society's curator of experiments.

His interest in microscopes led to the Society officially requesting him to arrange a series of microscopical demonstrations, and on 25 March 1663 this move was formally minuted. The following week plans were further defined, for Hooke was charged to bring in one new observation every meeting 'at least'. On 8 April he presented his first specimen – a tiny wall moss. One week later he demonstrated fine shavings from a bottle-cork, which showed how the properties of cork could be related to what the microscope revealed of its nature. Hooke explained that the nature of cork must be capable of explanation through its physical composition: it is light, does not absorb liquids, and it is compressible. The microscope gave him the answer.

> Why was it so light a body? My *Microscope* could presently inform me that here was the same reason evident that there is found for the lightness of froth, an empty Honeycomb, Wool, a Spunge, a Pumice-stone or the like; namely, a very small quantity of a solid body, extended into exceeding large dimensions.

So he proceeded through the structure of cork, diligently seeking a microscopical solution to familiar problems.

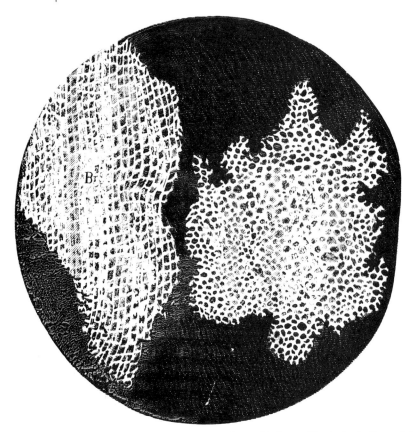

Fig. 3   The etchings of cork sections from Robert Hooke's influential *Micrographia* (1665). The appearance of this tissue led to Hooke's use of the term 'cell', which subsequently entered the vocabulary of science.

These pioneering observations with the compound microscope are to be found in the published pages of *Micrographia*. The capillary experiments that first brought him into the Royal Society appear on p. 10 of the book as 'Observations VI: Of small Glass Canes'; the studies of moss on p. 131, and of cork on p. 115. It was in the description of cork that Hooke gave us an important coinage which is part of modern biology, and familiar to us all:

> *Of the* Schematism *or* Texture *of* Cork, *and of the Cells and Pores of some other such frothy bodies.* [read the heading, and the text added] These pores, or cells, were not very deep, but consisted of a great many little Boxes . . .

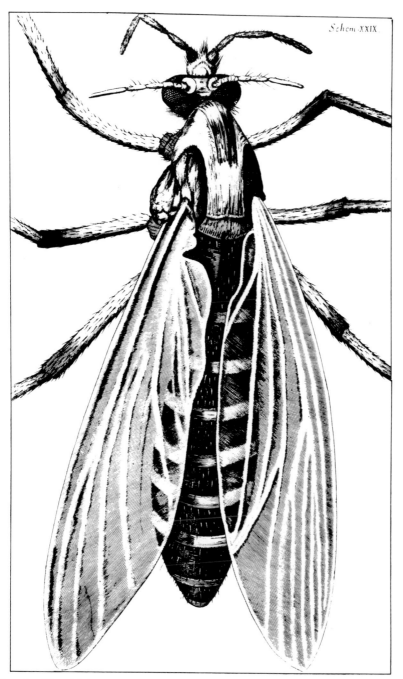

Fɪɢ. 4   The 29th plate from *Micrographia* showed Hooke's study of the 'Great Belly'd Gnat'. It was executed with care, though a comparison with Stelluti's figures of 1625 (p. 20) shows that Hooke was still using a magnification that had increased only a few times in 40 years.

It was in those words that the term 'cell' first entered the language of biology.

Micrographia became a great talking-point. It was published in 1665 and on 20 January of that year Samuel Pepys (who became acquainted with Hooke) bought a copy of the book and found it 'a most excellent piece, and of which I am very proud'. Pepys (who, like many men-about-town of the time, owned a compound microscope) sat up into the small hours when he took it to bed to read on the following evening, and was fascinated by it. Two years later, in 1667, the Royal Society issued a second edition. Facsimile editions have been reprinted since, three of them at least in the present century, and in an American reprint the book is still available to this day.

Hooke's book contained many fascinating illustrations, prepared from his own drawings and engraved under his own supervision. They are beautifully done, and to this day have a clarity and realism that few modern pictures could convey. He noted the 'cellular' structure of such specimens as elder-pith and the white of a quill pen, for instance; and he published some striking views of fabric seen under the microscope, the interwoven threads looking as bold as strands of rope. They were inspiring ideas, and eye-catching sights.

In due course they caught one eye that would alter the history of the microscope – and which gives the real starting-point for this book's exploration.

2 | The Draper's Question

One of the greatest mysteries about Antony van Leeuwenhoek is how you pronounce his name. There is even a degree of uncertainty about how you spell it. Leeuwenhoek himself spelt his own name in several different ways, and the English version of his forename is written *Antonj* in the Dutch original (-y being the English equivalent of the Dutch -j). In the United States there is a vogue for calling him 'Anton' but there is no precedent for that, apart from the use of that curious abbreviation by some German historians of previous generations. Leeuwenhoek's biographer Clifford Dobell, whose monumental volume *Antony van Leeuwenhoek and his 'Little Animals'* was published in 1932, to mark the 300th anniversary of his birth, noted nineteen different spellings of Leeuwenhoek's surname in the *Philosophical Transactions* of the Royal Society (in which the English translations of his many letters were first published), so it is highly surprising that – with such a range to choose from – the mis-spelling in vogue in the USA is one even Leeuwenhoek would not have recognised.

The pronunciation is not too difficult for an English-speaker to master. The forename is pronounced as it is in English, though with the stress on the middle syllable, rather than the first: it becomes An*tony*, rather than *An*tony. And the surname does not commence with 'Lieu-' or 'Low-' or any other of the popular choices, but with 'Lay-'. If you say it *Laywenhook* then you sound near enough to the right pronunciation from a Dutchman's point of view.

The other mystery concerns how he became interested in simple microscopes. That is, to me, a mystery no longer. But to see how the matter was solved, let us first sketch in something of his life story.

Antony van Leeuwenhoek was born on 24 October 1632 in the Dutch town of Delft and he died in the same town 90 years, 10 months and 2 days later (in the words of his epitaph in the Old Church), i.e. on 26 August 1723. His parents were Philips Antonyszoon van Leeuwenhoek, a basket-maker, and Margaretha Bel van den Berch, and they lived in the east end of Delft. He was christened 'Thonis' and his entry appears on the same page of the baptismal register in the archives of Delft as the great Flemish painter Jan Vermeer, whose executor Leeuwenhoek was to become after Vermeer's death in 1675 at the age of 43.

In 1638 the father died, and Leeuwenhoek's widowed mother remarried two years later, this time a painter, Jacob Molijn, who died eight years later. Leeuwenhoek's mother died in 1664 and was buried in the Old Church on 3 September. At the age of about seven, the young Antony Leeuwenhoek was sent away to school and went to live with an uncle at Benthuizen, a few miles north of Delft. And at sixteen, the year in which his mother's second husband died, he went to Amsterdam to learn the business of a linen-draper. During this time he may have begun his acquaintance with the young Jan Swammerdam, whose magnificent studies of insect life were to become internationally renowned.

But in 1654 he came back to his native town and he lived there for the rest of his life. He married Barbara, daughter of a Norwich cloth-merchant, at the New Church in Delft in 1654 – a year marked in the history of the town because of a massive explosion in the powder-magazine which, according to contemporaneous accounts, left virtually no house without some damage, and laid waste many streets; the death toll was never finalised. Five children were born of the marriage, only one of whom survived to adulthood. She, Maria, eventually became her father's housekeeper and assistant during the later years of his life. His first wife died in 1666 and in 1671 he married again, this time to a better-educated woman named Cornelia Swalmius (or van der Swalm). She bore one son, who died an infant, and she herself left Leeuwenhoek a widower in 1694.

Leeuwenhoek had bought a property in the Hippolytusbuurt in Delft, known as *Het Gouden Hoofd* – the Golden Head. Two bills he made out during this period have been discovered, which show the earliest-known examples of his handwriting and signature. At the early age of 27, he was nominated to the post of Chamberlain to the Sheriffs of Delft, a position that amounted to being a kind of civil caretaker: he was expected to be responsible for the security of the building, to keep fires lit when necessary, and to clean the rooms. The wage paid was 314 florins per year at first, which had risen to 400 florins by 1699.

It would be wrong to conclude from this that Leeuwenhoek was down on his knees, cleaning floors for a pittance. Dobell concludes that the position was more of a sinecure, the labour being performed by proxy, and that Leeuwenhoek was honoured by the appointment. It has been discovered that Leeuwenhoek was tested as a surveyor and geometry student in 1669, and in one of his letters (dated 14 March 1713) he wrote that he and another surveyor had both measured the height of the tower in the Delft New Church using trigonometry, each getting the same result (299 feet). The next documentary evidence of Leeuwenhoek's activities dates from 1676 when he was asked to sort out the financial problems left by the death of Vermeer, and in this way became the great painter's executor.

And that poses a puzzle. For it was in 1673, in this 'silent period' from the archives, that Leeuwenhoek suddenly emerges in the annals of London's prestigious Royal Society. The then young Society had been involved with microscopy for some time. The work of enthusiastic researchers like Wren and the successful publication of *Micrographia* by Hooke had attracted international interest, and in 1668 the *Philosophical Transactions* had published an account of a newly invented compound microscope made by Eustachio Divini in Italy with which, it was said, 'an animal lesser that any seen hitherto' had been discovered.

Five years later, the young Dutch anatomist Reinier de Graaf wrote a letter to the Society's secretary, Henry Oldenburg, about 'a certain most ingenious person here, named Leeuwenhoeck' who had devised microscopes that were far better than others 'made by Eustachio Divini and others'. De Graaf enclosed a short account from Leeuwenhoek describing his microscopical observations of a louse, the sting of a bee, and a mildew. These were translated, and duly appeared in *Philosophical Transactions*. Four months later, on 8 August 1673, Constantijn Huygens (the diplomat and poet father of Christiaan Huygens the astronomer) wrote a further letter of recommendation:

> Our honest citizen, Mr Leewenhoeck – or Leawenhook, according to your orthography – having desired me to peruse what he hath set down of his observations . . . I could not forebear to give you this character of the man, that he is a person unlearned both in sciences and languages, but of his own nature exceedingly curious and industrious . . . as you shall see by his cleere observations about the wonderfull and transparent *tubuli* appearing in all kinds of wood.

One week after that, Leeuwenhoek himself wrote his second letter to the Society. It set out his reasons for an insular life-style clearly enough:

> First, I have no style or pen with which to express my thoughts properly; secondly, because I have not been brought up to languages or learning, but only to business; and in the third place, because I do not gladly suffer contradiction or censure from others.

It was a frank and gutsy beginning to an open and straightforward relationship. Over the following years, until only a short while before his death, he sent 200 letters to the Royal Society, some of them illustrated, each of them of personal style, and the vast majority delving into new and mysterious matters which he systematically and intelligently explored. Leeuwenhoek discovered bacteria, and many other types of microbes too – his drawings of rotifers, beautifully-constructed pond organisms

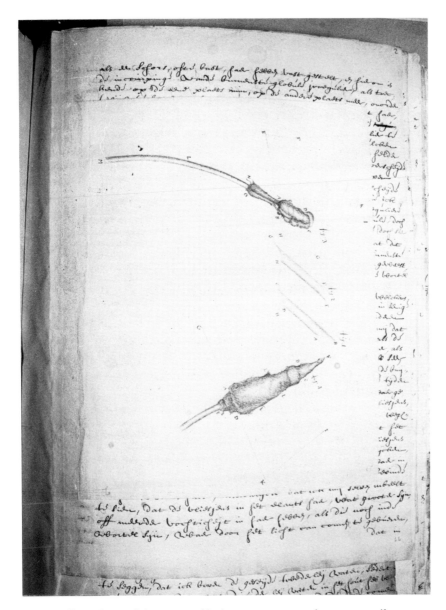

FIG. 5 Drawings of the roots of hairs are amongst the most easily recognisable of the figures accompanying Leeuwenhoek's letters. He did not make the diagrams himself, as he claimed to be bad at drawing. In each case they were executed by an artist.

with a wheel-like arrangement of microscopic cilia with which they sift their food from the surrounding water, were not to be exceeded in clarity for over a century – and he laid down many of the foundations of plant anatomy, the study of crystals, blood cells, spermatozoa and tiny creatures from lice to mites.

His reputation endured. As recently as 1862 it was being said that 'it is dangerous for any one, even now, to announce a discovery without having first consulted his works,' which gives him a longer run in the annals of current scientific research than almost anyone else. The *almost* is inserted there as a precaution against some wiser, professional historian citing an example; I am at a loss to find one myself.

Leeuwenhoek's microscopes were little slips of metal, perforated with a small hole, in which was sandwiched the hand-made lens. He must have made 500 or more in his lifetime, but only nine remain that are associated with him. The majority of his letters are preserved to this day in the vaults of the Royal Society, now in Carlton House Terrace. All of them have been carefully studied by historians and philologists, mostly working from microfilm copies of the delicate originals, even though English translations of many of them have still to be published.

What other relics remain? There are a few impressions of his seals in wax, including one with his portrait, another with a heraldic shield, and a third with a design of his initials. There are several portraits. The best known was by Johannes Verkolje (born in 1650) who lived in Delft from 1673 to 1693, when he died in the town. It shows Leeuwenhoek sitting at a table, on which lies the scroll of honour sent by the Royal Society to commemorate his election as a fellow on 8 February 1680. A more contentious example is the painting titled *The Geographer*, painted by Vermeer in 1669. This portrait shows a man who looks vaguely like Leeuwenhoek and near him stands a celestial globe much like the one in Verkolje's picture. The man holds a pair of compasses in his right hand too, just as the image in Verkolje's painting. The picture was even painted in the same year that Leeuwenhoek qualified as a surveyor, and it is really a surveyor, rather than a 'geographer', that the picture portrays.

Leeuwenhoek also appears in a group study by Cornelius de Man, and there is a little oval portrait on the title-page of the

FIG. 6   Leeuwenhoek at his desk. This engraving appeared as the frontis-piece to a nineteenth-century biography, and was taken from the oil portrait by Verkolje which hangs in the Leiden museum.

published versions of Leeuwenhoek's last letters. Original artwork for just such a design was rediscovered in the early 1920s with Leeuwenhoek's own signature on the paper, dated 30 April 1698; and this is certainly the only 'autographed picture' of its kind to have survived.

What of Leeuwenhoek's dwellings? There is a plaque set into the wall of a shop that now stands on the site of Leeuwenhoek's house, where he made those great discoveries. The premises are now run by the Dutch firm of V & D, and so it remains, by some curious chance, to this day a draper's. The

house where Leeuwenhoek was born was painstakingly identified by L. G. N. Bouricius, a devoted scholar of his scientific
work, in the early 1920s. He published an account of this
sensational revelation in *De Fabrieksbode* of March 1925.
Dobell, much excited by the news, went to Delft to take his own
photograph of the building in the following year. Then, two
unfortunate mishaps occurred: in the first place, Dobell's
'exclusive' pictures were published before he had a chance to
record them in the literature of science, and secondly – far
worse! – the house was shortly afterwards demolished.

So there were two tantalising areas left to explore in
studying the work of this pioneering Dutch draper. One of
these was the vexed question: 'where did he obtain his inspiration?' There were many answers to that, but in time I found
they were wrong. Even the judicious summing-up by Clifford
Dobell proved to be in error. The second puzzle was the most
tantalising of all – namely, were there any other relics of his
work still in existence? Was there a source of first-hand information on the way he had worked? Could it be possible that
there was something from Leeuwenhoek himself, waiting to be
investigated, that could throw new light on his work and its
motives?

The answer to the second question was, much to my great
surprise, 'yes'. And from a study of that question, an answer to
the first would arise.

3 | A Scientist's Legacy

I first came across the name of Leeuwenhoek when I was fourteen or fifteen, at about nine o'clock on a Thursday evening. It was then I used to have regular tutorials with Dr A. G. Lowndes, a knowledgeable and stimulating yet intimidating biologist from the Marine Biological Research Station at Plymouth. Quite how he came to be a tutor at my school was a matter I never resolved, for he was a research biologist before and after. I imagine that his enthusiasm for offering to inspire a schoolboy scientist with a love of biology was more to avoid living every night surrounded by boarders in School House rather than out of pure altruism. He used to divide his time between giving me tasks to do – like memorising the elements and their symbols, which I later found useful – and talking about ideas in microbiology. He was the first person to allow me access to a fine research microscope, and I have never doubted that his influence was salutary.

During one of those tutorials he mentioned the name of Leeuwenhoek. The name stuck in my mind simply because it was unusual, and I later found it in an old and sagging book on bacteriology that I used to borrow from the library. The first time I began to find out some details of Leeuwenhoek's work came one spring day when I was nineteen. I used to spend much of my spare time as a rhythm and blues pianist with a rock band, and I turned up one day too early for practice, and found the door of the church hall still locked. To kill time I browsed round a neighbouring bookshop and it was there I chanced upon a book called *The Microbe Hunters*. It was the same book that gave rise to a BBC tv series of drama documentaries of the same title in the 1970s. The author was an admirable American named Paul de Kruif, and the first chapter in his volume of biographical essays was devoted to the man by name of Leeuwenhoek, that had so often cropped up in the past. I kept reading the book during band practice later that day, and even announced with a flourish that my next number would be a nifty little stomper called 'Leeuwenhoek Blues'. The suggestion met with the response you might expect for a stray dog in the school yard: a subtle blend of hysteria and antipathy. I soon found that most people knew nothing of the man, and in the length and breadth of the University Library I never found a description of his life, or his work.

My acquaintance with Leeuwenhoek came about gradually after that. I traced the books by Schierbeek and Rooseboom, Dutch historians and scientists who had spent many years sifting through the evidence of the man's life. Maria Rooseboom and I corresponded briefly, and at one time I had a three-quarters-formed plan for us to meet in Holland, but that came to naught before she died. In due course I came across a library copy of Dobell's detailed book on Leeuwenhoek's microbiological researches, which tells so much of the other aspects of his life, and found that a most thorough and detailed book, justly famed in the history of science.

Leeuwenhoek tended to crop up from time to time, and on each occasion added more to my unravelling understanding of the way biology had developed. One of the first practical investigations I set myself was the problem of how you could see so much detail with nothing more refined than a single lens. Practical experiment was the answer here, of course. I tried using small single lenses from modern microscope objectives, with fair results; but they did not approximate to the results obtained using a primitive hand-made lens from three centuries earlier!

To overcome those objections I made some imitation

simple microscopes myself, with tiny beads of unpolished glass that I produced in a bunsen burner flame for the purpose. The results appeared in the photographic journals, and shortly afterwards in the magazine *New Scientist*. I had shown that you could obtain surprisingly good results with such simple apparatus, and the correspondence that followed showed how intrigued other people were by the results. Using conventional light it was difficult to see details in specimens prepared for the microscopes, but with a narrower beam of light (in some of the experiments we used sunlight) it was possible to obtain clear studies of plant cells that belied the simplicity of the magnifying element. Several of the pictures were reproduced abroad – a new teaching course in Australia used the results of the experiments in setting up a study for schoolchildren, which was greatly encouraging.

It is fair to say that these experiments did *not* convince me that Leeuwenhoek could have seen all he claimed. My feeling at the time was that he may have used some now-forgotten method to obtain his results. We are all children of our age, and we react according to the values and the criteria we learn from our peers and our experiences, and I felt so convinced that simple lens systems would not be up to the mark that I speculated whether Leeuwenhoek had secretly used two lenses together. Do you see why? It was simply that my own understanding of modern microscopy was so heavily weighted against the single-lensed microscope that it did not enter my head they could be as good as they needed to be. I had been brought up on compound microscopes, and was so used to the implied 'fact' that they were bound to be vastly improved over the simple microscopes that Leeuwenhoek used, that it was a perpetual temptation to conclude that his answer had lain in imitating a compound instrument in his own primitive way.

Leeuwenhoek added to the problem by being deliberately secretive about his methods. He would never let people from outside his own closed circle look through his best instruments; he often said he had special methods for highest magnifications; and all of his occasionally almost petulant comments in this vein add up to a determination to hide his ways of working. So the fact that he might be using two lenses together – an inevitable consequence of the massive propaganda lobby in favour of compound microscopes – was not altogether an unreasonable conclusion.

But this was not the answer. When I spent time looking back into the earlier years of this century I came across others who had tried to imitate Leeuwenhoek's work. One of the most successful attempts was the work of J. Kingma Boltjes in the 1930s, who had some tiny lenses ground in imitation of those made by Leeuwenhoek. He brought into focus bacteria, protozoa, and a whole range of microscopical objects for scrutiny, and even obtained some photographs. What these elegant and convincing experiments had shown was, without doubt, that Leeuwenhoek microscopes *could* see extraordinarily fine detail. It was not necessary to rely on the compound system of lenses at all. Dobell's work had done little to elaborate this point, because – surprising as it seems – he did not find an opportunity to examine any of Leeuwenhoek's lenses. Apart from two Leeuwenhoek microscopes in cases at Leiden, Dobell only inspected one of the instruments closely, and that was the property at the time of Miss S. A. E. Haaxman, an indirect descendant of one of Leeuwenhoek's nieces.

Clifford Dobell made a serious mistake when he performed his only recorded experiment with a single lens (p. 330 of his famous book on Leeuwenhoek). He used the front lens of a modern microscope lens, a highly corrected 2 mm apochromat, without restricting its aperture. Leeuwenhoek himself knew that the lens should be stopped down, and also that the cone of light illuminating the specimen should be restricted – his advice was to use light from a window, since he knew that a small light-source gave a clearer image. Dobell used a clear sky as his own illuminant, and this would have given him an image of exceedingly low contrast. It was remarkable that he saw anything at all. This experiment explains why Dobell was unable to repeat Leeuwenhoek's experiments; and it also accounts for the emphasis he later placed on his search for the ways Leeuwenhoek might have used to produce images of higher contrast.

Photographs of the actual images generated by the Leeuwenhoek microscopes were few and far between, and although there had been some tests carried out there was still no thorough examination of the optical performance of all the Leeuwenhoek lenses still in existence. Many historians had raised doubts about the ability of the microscopes to demonstrate as much as was claimed. One ambition did form in my mind, and that was to carry out some detailed studies on the lenses in order to prove the extent of their capability.

Meanwhile a different conundrum had continued – the simple question: *where did Leeuwenhoek get his initial inspiration?* The books seemed reasonably unanimous on the subject: he had trained as a draper, had used lenses to check the quality of cloth, and had from that practical experience launched himself into the realms of high-powered microscopy. Does that seem likely? To my mind there was a considerable leap in the middle of that argument, and the more I thought about it the less probable it appeared. Was it possible that Leeuwenhoek had been actually taught about microscopes by someone else? That was a seemingly more sensible argument to consider in detail. So let us examine the candidates.

Leeuwenhoek knew Jan Swammerdam and they may even have met first when he was in Amsterdam (p. 26). Swammerdam became one of the great pioneers of insect micro-anatomy. Certainly he used microscopes in his work, single lenses made by Musschenbroek for low-power work, and his own blown lenses for higher magnifications. Here would be a clear possibility – until we look closer at the dates. Swammerdam was 11 when Leeuwenhoek went to Amsterdam, and only about 16 when he left. The chances for any deep-seated academic stimulus from that source would appear to be slight – Leeuwenhoek was five years older, and it is hard to imagine a 21-year-old learning much from someone of 16. Not only that, there is no evidence that Leeuwenhoek had the least interest in microscopy for some twenty years after that time. Swammerdam, a promising candidate at first glance, may be taken off the shortlist.

Then there was Nicholas Hartsoeker, an astronomer-mathematician who knew much about lenses. Hartsoeker certainly knew Leeuwenhoek, and wrote about his visits to the house where he worked. He also plagiarised Leeuwenhoek's discoveries and tried to present them in foreign towns as his own. Leeuwenhoek wrote about Hartsoeker, sometimes in less than polite terms; indeed his last comment on him reads: 'It has come to my ears that Hartsoeker has not much of a reputation among the learned, and when I saw that he laid claim to untruths, and was stuck-up, I looked into his writings no further.' Perhaps Hartsoeker's greatest spurious claim was to the discovery of spermatozoa, which he presented in Paris as a piece of original work after he had been shown them by Leeuwenhoek. But might he have been a source of inspiration in earlier days? Here too the dates do not match, for Hartsoeker was born in 1656, according to his own writings, and was therefore only 17 when Leeuwenhoek first communicated with the Royal Society. At that time, he would have been able to offer little by way of encouragement, for Leeuwenhoek was some quarter-century older and Hartsoeker's work was yet to begin.

Then there are Leeuwenhoek's two sponsors who introduced him to the Royal Society: de Graaf and Huygens. The tone of their letters shows that they were helping a newcomer to the field, and were not in any way Leeuwenhoek's mentors. De Graaf, remember, wrote of 'a certain most ingenious person here, named Leewenhoeck, [who] has devised microscopes . . .' and Huygens added, 'Mr Leeuwenhoek . . . has desired me to peruse what he hath set down of his observations.' Neither man was implying, or even admitting the vaguest possibility, that they had taught Leeuwenhoek microscopy. There are suggestions that Leeuwenhoek had learned from the Italians (such as Divini) who had already established reputations for microscopical investigation. A letter discovered amongst the Magliabechi manuscripts in Florence in 1929 was claimed to have originated from Leeuwenhoek on 2 May 1692. Dobell found that the letter was certainly not written by Leeuwenhoek at all, but by G. W. Leibniz. Firstly, the letter was in Latin (a language never used by Leeuwenhoek, and which he could not speak). Might Leeuwenhoek have had the letter translated by someone close to him? Certainly – but on closer scrutiny this letter turns out to have been written from Hanover, it bears Leibniz's private seal, and it was originally published as a genuine Leibniz letter in 1746.

Clifford Dobell was quite firm in his own conclusion as to Leeuwenhoek's introduction to microscopy. He was self-taught, says Dobell, and that's that. In one of Leeuwenhoek's earlier letters is a reference to a visit he had paid some years earlier to London. The paragraph comes in his letter No 6, dated 7 September 1674:

About six years ago, being in England, out of curiosity, and seeing the great chalk cliffs and chalky lands at Gravesend and Rochester, it oft-times set me thinking; and at the same time I tried to penetrate the parts of the chalk. At last I observed that chalk consists of very small transparent particles; and these transparent particles lying one upon another is, I now think, the reason why chalk is white.

Dobell writes that it is clear Leeuwenhoek had a holiday in England in 1668, and arrived with a microscope in his possession. He concludes that Leeuwenhoek was 'certain' to have begun his microscopic studies during the life of his first wife, i.e. prior to 1666.

In my view these are doubtful conclusions. It is not 'evident' that Leeuwenhoek had a microscope with him on the visit. His use of the expression *te penetreren* means just what it says – that he tried to penetrate the parts of the chalk, as he termed them. He may have dug at it with a knife-blade, or even attempted to work out how the chalk was composed; but nowhere in his words did Leeuwenhoek imply that he did anything microscopic. It is not even sound to conclude so assertively that the visit must have taken place in 1668, for Leeuwenhoek says '*about* six years' before 1674, and that could just have well been late in 1667 . . . the year in which the much-discussed *Micrographia* of Robert Hooke was published in its second edition, just when the British–Dutch naval conflict was dying down.

That seems to me a far more likely explanation. We have to make some far less extreme assumptions to propose that it was from Hooke's great work that Leeuwenhoek received the trigger for his own life-time of study and experiment. The fact that he was so interested in the structure of chalk, and deduced that its colour was due to the presence of transparent particles overlying each other, like snow, shows he had an inquisitive mind. If he really had used a microscope, he would never have drawn these conclusions, for he would have seen that the chalk was composed of easily-recognisable minute foraminifera, their remains looking like microscopic sea-shells. Even a low-power lens reveals that.

What would happen to such an inquisitive person, naturally interested in the structure of materials, arriving in London around 1667 or 1668? His turn of mind would be apparent to his hosts, and indeed if Leeuwenhoek visited friends they would very likely have been of a similar temperament themselves. The chance that they would have shown him London's literary and scientific talking-point, seems obvious. For Leeuwenhoek there was an even greater motive – he was a draper, he was doubtless meeting drapers in his travels, and in Hooke's illustrations were several fascinating and revealing studies of fabrics, including a delicately drawn portrayal of what we now know as 'shot silk' (i.e. a shiny form of material in which a reflective sheen, often of two colours, is visible). The very fact that the 'book of the moment' contained examples of the materials with which Leeuwenhoek worked makes it, to my mind, overwhelmingly likely that he was shown *Micrographia* – and was captivated by its message.

The reasoning does not end there. One important passage of Hooke's Preface holds a significant clue, for in it he mentions a different kind of microscope which he prefers not to use:

> If one of these lenses [a drawn-out thread of glass run into a bead and polished with jeweller's polish, just as Leeuwenhoek used to do] be fixed with a little wax against a needle hole pricked through a thin plate of Brass, Lead, Pewter or any other metal, and an object be placed very near be looked at through it, it will both magnify and make some objects more distinct than any of the large microscopes. But though these are exceedingly easily made, they are yet troublesome to use.

Here we have it: an exact description of the kind of microscope Leeuwenhoek perfected. For the Dutchman, the 'difficulty of use' was not the overriding consideration; simplicity of construction was. In Hooke's case it was quite the converse, for he had to carry out his many duties as an agent of the Royal Society and the cost of a microscope was a lesser difficulty – if he wanted an instrument, he would doubtless send out for one to Christopher Cock or one of the other microscope manufacturers popular at that time.

It was in the closing weeks of 1980 that I sought the help of the Royal Society in commencing a systematic review of the surviving literature on the topic. In February 1981 I had collated large numbers of original documents on the subject, and it was then I decided to look at the original correspondence of Antony van Leeuwenhoek, which lies to this day in the double-locked vaults of the Society's premises. My interest in looking through the letters was understood well enough, though I think it was felt that there was little to be gained. Scholars in the past had sifted their meaning time and time again, and microfilm copies of the letters had been analysed by translators in Holland, and there was under way a continuing programme of publication for all the letters – a task that had started in 1939 and was half-way completed by 1981.

Often when working with source documents I have been diverted by the aroma of newly opened files of letters, for the smells of an earlier age emerge with the dust. You can smell the charred tobacco scent of clay pipes, and the sooty scent of smoking chimneys and log fires. I have often thought that there should be a discipline known as 'Olfactory archaeology', in which one would analyse those evocative smells and relate them to the life-style of the time. For any practising microscopist, the opportunity to look closely at some of the specks of dust or a shed whisker or two in Leeuwenhoek's letters that might have held some fresh clue about his life, or his methods of working, was very enticing. Besides which, I had in my memory the fact that his letters had – more than once – referred to material he was going to send to the Society. Surely there might be *something* worth looking for? I was asked at the time if I had any preconceptions as to what might be there, and I said, 'Perhaps he snipped some hairs from his wig, and left those,' for he used the thickness of a hair as a reference object in calculating microscopic dimensions.*

What I actually found exceeded anyone's most exagger-

---

* Some confusion was caused for many years by the fact that the objects Leeuwenhoek measured by this means seemed to be too great. If his measurements were accurate – and all the others he made were noted for this – then a human hair would have measured only about 43 μm (i.e. 43 thousands of a millimeter) in thickness. Actually, a human hair is more like 70 μm in diameter. It later was realised that Leeuwenhoek used as his reference point a hair *from his wig*. The kind of wool used for wig-making at the time comprised hairs that were a good deal finer than human hair – and 43 μm turns out to be exactly the right dimension.

ated aspirations. Leeuwenhoek's letters are on hand-made paper, written in ink that is dark brown in colour rather than black, in a hand that is at first sight hard to read. Slowly I worked through the text of the first few letters, relating the decipherable terms to what I knew of the published translations as I went. There was one immediate disappointment, for the manuscript of his first-ever letter – which had been transmitted to the Society by de Graaf – was by then missing. In it, Leeuwenhoek had taken to task some unnamed observer of the sting and mouth-parts of a bee, a louse, and the fruiting structures of a mould. Leeuwenhoek was surprised that this anonymous observer had failed to see the seminal property – spores, or seeds, if you prefer – of the mould specimen, and pointed out that they were certainly real, even if his 'rival' had missed them. This was a hint at a continuing belief of Leeuwenhoek's, namely that spontaneous generation was a myth. He took exception to people who implied (as was the case here) that a mildew could propagate itself spontaneously, without any spores.

The target of this attack was plainly Hooke, for it was he who had made exactly these suggestions in *Micrographia*. Leeuwenhoek wanted to prove him wrong but – in the style of the times – he did not stoop to mentioning his name. A sight of that historic, first-ever document would have been intriguing. Though not well-written, it has an important niche in the history of microbiology. I leafed on through the letters, turning the pages carefully and taking the chance to inspect them closely as I went. And then occurred one of the most vividly remembered moments of my life. As I lifted the final leaf of Leeuwenhoek's letter of 1 June 1674, it felt heavy. Pasted to the back of the last blank page was a white paper envelope. It was square, about the shape of a greetings-card envelope, and the rectangular flap was tucked inside. Carefully I opened the flap, and removed what lay within the envelope. There were four small square packets made of folded notepaper – and each was annotated in Leeuwenhoek's own distinctive handwriting. Using forceps and instruments, I cautiously opened the packets and looked inside. I did not breathe; my concern being to avoid contaminating by breath whatever lay within.

And there lay Leeuwenhoek's own original handiwork. I could see what several of the specimens were, and it was

FIG. 7   The packets of specimens at the moment of discovery in 1981: 'pith from elder' is at the top, with 'Cork' beneath. The letter with which they were sent was dated 1674, before Leeuwenhoek acquired the 'van' in his signature.

possible to relate them to what was written on the packets. The first was a set of beautifully fine sections of cork. The legend on the packet confirmed that: 'Kurck' it read. In the next packet were glistening-white sections of elder-pith, known to every student of microscopy as a favourite supporting medium when cutting materials (like sections of leaf) that are otherwise weak and flexible. 'Pit van Vlier', read the label – pith of elder. A third packet was empty. Just a few traces of dusty matter remained. And in the fourth and final paper container were some rounded, thicker sections. I could not immediately make out the wording of the long caption Leeuwenhoek had written on the outside, nor could I positively identify the contents. Each time I needed to breathe while this was going on, I gently closed the packets and turned away. My camera was standing on its tripod alongside, and I tripped the shutter mechanism and the autowind as this exploration process was going on. I only had a medium-speed colour transparency film, for my intention had been to take some photographs of the hand-written pages of the letters themselves. I resolved to sprint up the hill towards Piccadilly Circus to the chemist's for a supply of black and white film, so that there would be some printable negatives of this surprising discovery.

I was beginning to run short of printable language too, by this time, and was rescued by Professor Allibone, a fellow of the society, who came down into the vaults to find me, alone, surrounded by the paraphernalia of investigation. 'What you need, after a moment like that,' he sensibly proposed, 'is a cup of tea. Come along and tell me what you've found.' In the interests of objectivity I have to say that tea is something I rarely drink in England, it being one of the few countries where a well-made cup is – for my taste – more than elusive. That cup, that afternoon, was better than a glass of Armagnac and a Havana cigar.

Over the following days I searched methodically through Leeuwenhoek's remaining letters. In all, nine packets came to light. Two of them still held fragments of cotton seeds, which had been carefully dissected open to reveal their internal structures. Three others held dried specimens of pond mic-robes, mostly algae, which Leeuwenhoek had described in 1686. What a selection! It was not as though I was presented with nine similar specimens – nine kinds of seed, say, or nine types of leaf – but instead there were fine sections, neat dissections, dried microbial material . . . a veritable store-house of differing microscopical techniques, preserved over three centuries for scrutiny in our own era. An immediate coincidence drew itself to my attention – we had nine micro-scopes associated with Leeuwenhoek, and here were exactly the same number of specimen packets.

In some ways the packets were the more intriguing. They were annotated in unmistakeable handwriting, for example, whereas the evidence linking the microscopes with Leeuwenhoek is in most cases indirect (and in at least one case virtually non-existent). A second point is that the specimens were likely to be described in the letters that had accompanied them, which did not apply to microscopes which happened to have survived. Thirdly, the materials themselves were pre-pared by Leeuwenhoek's own 'mysterious' techniques. That meant that one could try to re-examine the specimens and relate them to his own contemporaneous descriptions; it allowed one to feel that – at last – it might be possible to identify the exact nature of the types of organism he described, since in previous investigations that had to be done by inference and guesswork; it gave first-hand evidence of his techniques, and showed how carefully he could handle material (a unique and important benefit to any investigator); and of course there was for me the added possibility that one might find out some hidden secrets by closely examining the specimens. Microscopic traces left on them could be traced back to facets of Leeuwenhoek's own time, and perhaps to the man himself.

It was not enough merely to consider the specimens themselves, but to relate them to their context in history. One of the most helpful comments I received from Sir William Paton of the Society was: 'The important task is to get as much information from the specimens as possible,' and this was what I set out to do. Analytical 'detective' microscopy is not a widely taught activity these days. There are individuals who work with scanning microscopes, others who are experienced microanalysts, plenty who use the conventional transmission electron microscope, a good number of enthusiastic optical microscopists who use the well-established methods of routine light microscopy. Clearly what these specimens needed was investigation through a pooling of methodologies.

First things first. I was allowed by the Society's archivist to take portions of the specimens away for examination. Everyone was astonished at the find, and I removed only about one third of the material in each case so that the rest could be kept in original condition. In a few decades' time it might be possible to culture viruses from dried specimens of that kind, though we do not have suitable techniques at present to do such a project justice. I did suggest that the specimens might be removed from the letter files and kept away from any chance of outside contamination, and that was agreed. When the time came for the packets to be removed, some of them could not at first be found amongst the letters – little wonder they were overlooked so often in the past!

The first aim was to look closely at the specimens, and so I prepared a small sample of each for examination in the scanning electron microscope. The instrument in the Zoology

Department at University College Cardiff was made available for my work and I shared several long and animated conversations about the specimens with the Professor of Zoology, Denis Bellamy, who was as excited about them as I was. Everyone wanted to know just what one would find.

The scanning electron microscope produces an image on a kind of small television screen. A beam of electrons scans across the specimen, and the reflected electrons are picked up and reconstituted into a solid-looking image, full of crisp detail and sharp reliefs. There is a tremendous depth of field to a scanning microscope image, so that a remarkably realistic impression is conveyed of how a specimen actually *looks*. The object itself has to be made electrically conducting for this to work, and one successful way of ensuring that is the case is to coat the specimens with a fine layer of gold. This metal is so resistant to tarnishing, even over a prolonged period, that a gold-coated specimen remains in viewable condition for years. If a speck of dust gains access to it, you tend to see nothing since the intruder is not gold-coated and so does not show up in the same way.

The first sliver of cork from the hand of Leeuwenhoek slid into the specimen chamber of the Cambridge Stereoscan 600, and the image began to form. There was the sight of a life-time – a delicate, magnificently controlled hand-cut section under a modern microscope after 307 years in the dark. Immediately some conclusions could be drawn. Older historians had written of Leeuwenhoek 'tearing apart' plant tissues to look at them better; even the most modern history of the subject dismissed early specimens as being prepared without 'finesse'. Without any doubt these sections were as good as the best hand-cut section might be today. The finest parts were only a few micrometres (thousandths of a millimetre) thick, and the detailed structure of the cell walls of the cork tissue glowed beautifully from the cathode-ray screen of the microscope. 'One would scarcely think Leeuwenhoek a major figure in botanical microscopy,' wrote one recent account. Leeuwenhoek was a dilettante; he should be remembered as a 'microscopologist rather than as a microscopist' . . .

How wrong! In the earliest years of microscopy, when Leeuwenhoek was learning how to cut his own material with a sharp shaving razor (as he himself notes), he obtained results that would do no little credit to a modern-day biologist. It was in cork tissues, as we saw on p. 22, that Hooke had first coined the fundamental biological term *cell* about a decade before this perfect little section had been cut. Here was a sight that took one back in time to the origins; here, before our eyes, were the roots of present-day biology. History was recreated in more ways than one: the modern botanist uses a mechanical microtome to cut fine sections, having first embedded the specimen in a supporting medium (usually plastic). But in the field, when a microtome might not be so readily available, we use a cut-throat razor to this day for the preparation of emergency sections.

I spent some time searching through this section of cork, and learned how Leeuwenhoek cut his sections to give them strength; how well he had sharpened his blade (for the cut surface was astonishingly regular); and how fine were the finest-cut cells he could obtain. Here and there were spores and pollen-grains which could provide extra evidence about the conditions at the time the sections were cut, and the way they had been stored meanwhile. Many of the images of those particles remain to be analysed, but they will throw further light on seventeenth-century science.

Next were the elder-pith sections. The cells of the pith that occupies the centre of an elder stem are larger than the cells of cork, indeed they are just visible to the naked eye. You can take – as Leeuwenhoek did in May 1674 – a two-year-old stem of an elder bush, cut a short length, trim away the bark and see the tiny glistening cells if you look sufficiently closely. The section I chose for examination in the scanning microscope was equally delicate and just as carefully prepared as the cork specimen. The cells stood out clearly, their polygonal walls crisply preserved and well marked. Here and there fine hyphae from a germinated fungus spore snaked across the tissue, though damage like this was slight. Because of the greater size of the elder cells there were comparatively large areas of flat cell wall to examine in greater detail, and I searched through the tissues paying special attention to the tiniest particles. Here and there were pollen-grains, just as before; but in one corner of a cell I saw something that had an unmistakeably familiar look to it. It was the round, concave shape of a human red blood corpuscle – an erythrocyte, or red cell as we more usually call them.

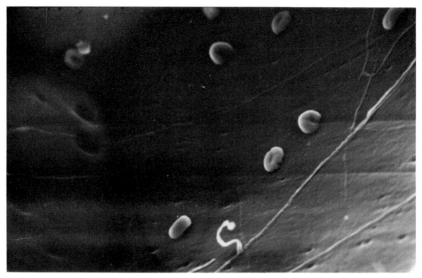

FIG. 8   Perhaps the most personal legacy: red blood cells (erythrocytes) remain on the surface of a section cut by Leeuwenhoek in 1674. He used an open razor for this work, and they were probably deposited directly from the blade.

Nearby were several others. On closer inspection the cell-walls could be seen to have become eroded over the centuries – testimony to their age – but the familiar shape of a red cell was unambiguous in appearance. Nobody can shave with an open razor without leaving behind red cells and yet the delicate cell-walls burst in an instant if they are brought into contact with fresh water, a process known as lysis. None of these cells had lysed, and so you might deduce that when Leeuwenhoek shaved he did so with a dry blade.

Or did the cells get on to the sections because of coughing impact? That is another alternative and, though I doubt whether it applied to the red cells, it certainly seems likely to have been responsible for another group of cells which I eventually found. Here what caught my attention was a round, dome-like inclusion, not in the least like a plant cell – a spore, for instance – but typical in appearance of a dried animal cell. Scattered around it were numerous minute, rounded objects, each one about 2 micrometres (or rather less) in diameter. Under the highest power of the microscope they could be seen to show the typical appearance of rounded bacteria, like staphylococci, and some could be seen in the stages of dividing in two: the elongation, followed by the 'nipping' in at the centre which allows one cell to form two. The larger cell, on the other hand, was the right size and shape to be a leucocyte – a white cell – from human origins. Here was evidence that made one wonder whether Leeuwenhoek had a sore throat on the day in question (which is possible; sore throats are rather more common in early summer). He certainly wrote on more than one occasion that he was liable to bronchitis, and so he is known to have had some bacterial involvement of his upper respiratory tract. Here were signs that you might expect to find in a modern clinical investigation, if it took your fancy to seek for bacteria and leucocytes with a scanning microscope; but these were ancient, eroded structures – the legacy of a long-dead pioneer who little knew what he might bequeath to an earnest follower three centuries into the future.

The third packet was empty; the legend said it had contained the 'white from a quill', and if any of these historic specimens had to be missing this was the one to dispense with. Feathers grow as non-cellular structures, like finger-nails or hair, and the 'white' from inside the feather shaft is not of great microscopical interest. It would have been good to know that all the specimens were intact, but modern investigations could not learn much from that source of material.

In the fourth packet was a series of small, neat, rounded slices. 'Sections' would be too good a word for them, as they were too thick to amount to a *section* in the microscopic sense. Leeuwenhoek knew this: he termed them 'slices' instead, which is exactly right. They were pieces cut from the optic nerve of a cow. Leeuwenhoek often wrote of the optic nerve, for it was intriguing to the pioneers to try to envisage how an image could pass from the eye into the brain, and clearly the optic nerve was the site of this miraculous process. Leeuwenhoek did not have any means of embedding the tissue in a supportive medium to make it easier to cut, as we would do today; instead he allowed fleshy animal tissues to go dry and hard, and then sliced them up with his razor. This means that the optic nerve slices are not very fine – the thinnest is about one-fifth of a millimetre – that is to say, 200 μm thick. That is ten times thicker than the sections of plant material.

Leeuwenhoek's interpretation of what he saw was clear enough. He knew that the open spaces in the nerve cord (where the tiny bundles of nerve fibres once were) resulted from material drying up and dropping away as the slices were prepared. He observed that the outer layer of the nerve cord was in two parts, and this is a peculiarity of optic nerve alone. As it happens, Leeuwenhoek was a poor artist, and he never did any drawings as far as we know. Instead, he used to show what he was observing to an artist, and would then check over the results. It is a strange way of working, but seems to have been very successful. There are two drawings of optic nerve in existence from Leeuwenhoek's desk which show the amount of detail he could observe. One of these is attached to a letter in the Royal Society files, whilst I found another unpublished original in the collection of manuscripts at the British Museum.

Apart from the nerve tissue when this section came in for scrutiny in the scanning microscope, I found some little pieces of mites here and there – and in one corner, the recognisable trunk of a mite. Did this indicate that the specimens had been attacked by mite during the centuries of storage? I think not, and this for two reasons:

First, there were no signs of the droppings or other evidence of activity – the paper wrappers had not been attacked, for instance.

Second, Leeuwenhoek himself noted that his dried animal specimens were already infested with mites before he prepared the material for study. These were mites that he knew to be present before the specimens were sent to London, and they provide an added source of information. Leeuwenhoek had prepared drawings of mites, he wrote about their life-cycle, he studied their breeding habits; and from many of his accounts it had been possible to work out what the likely species were. Here the guesswork ended – I now had photographs of the very mites from Leeuwenhoek's house in Delft of the seventeenth century, and we could obtain from them first-hand evidence of identification.

So far I had only considered the specimens from Leeuwenhoek's letter of 1 June 1674, and a wealth of new information had been obtained. The next letter with which specimens had been associated was dated 2 April 1686, and described his work on the structure of cotton seeds. This is what the two packets contained: one of them contained a cotton seed that had been cut into 24 round slices – 'ronde schijfjens' in his own handwritten caption – whilst the other contained portions of nine seeds that had been cut open, and their covering layers removed. These specimens had been moist when prepared, since they were fresh specimens. They were each rather overgrown with fungal hyphae and fine details of the specimens themselves were hard to make out. But the sectioned seeds revealed the details of the cotton embryo that Leeuwenhoek wished to investigate, and there was one surviving example of a neatly opened seed, the tiny root – the radicle – displayed, and the embryonic first leaves laid out. A drawing Leeuwenhoek sent to the Royal Society shows a similar example, and of course this might be the identical seed.

The remaining three specimens come from his letter dated 17 October 1687. In each case they take the form of flat, dry fragments of material from water: algae, and microscopic animal life all in a confused and brittle mass. To the naked eye such material offers little to entice the imagination, and under the scanning microscope the results, though an improvement, are not truly very revealing.

But these specimens tell an interesting story. In the seventeenth century it was believed that occasional scraps of paper fell from the heavens. By the time they reached earth, it was said, they were blackened. From time to time people claimed to find examples of this 'heavenly paper' and the legend that these were celestial missives (which had presumably been charred by the heat of re-entry) persisted. One such sighting had occurred in the State of Courland, on the shores of the Baltic, and the finder of the paper had sent a small portion of it to Leeuwenhoek for his interest. The caption on the scrap of paper in which the specimen was wrapped is written in an untidy hand, clearly not Leeuwenhoek's, and obviously by someone unused to Dutch. Some of the letters had been altered, and the date – which literally reads 'the 24 or 15 March' – is clearly meant to read '14 or 15' instead.

The specimen looks like charred paper. It is black, lustreless, brittle, and the right thickness for paper. But in fact the material is a dried sheet of algae, the greenish growth (one example of which children in many parts still call *witches' hair*) found in still water and, occasionally, in streams. What did Leeuwenhoek conclude? His words show the mettle of the man:

I had not had this supposed paper in my house for half an hour before I had (with the aid of the microscope) formed such a clear idea of it, that I fancied it to be a plant which had come forth from the water. I . . . believe that, owing to heavy rains or melting snow, the water from a morass or from ditches has flooded some piece of land, and that the water has left behind this green plant, from which the so-called paper is made, and . . . the sun and the wind caused the plant to become dry and stiff, so that it took on to some extent the look of burned paper.

More recent research workers have written of blackened sheets of papery material left behind when stranded mats of algae are left to dry in the sun, and the phenomenon is typical of the kind of marshy ground in the area where this specimen originated, now part of Latvia. Under the scanning microscope the 'heavenly paper' looks, none the less, remarkably like paper. One of the people who first saw the images had experience of working with paper samples in exactly this way, and was inclined to agree with Leeuwenhoek's unnamed correspondent that 'paper' was the identity of the mystery material. Closer examination, however, proved the watery origins of the material, for occasional recognisable cells of aquatic microbes could be discerned. It is a tribute to Leeuwenhoek that he found the answer as surely and expertly as he did.

Faced with this specimen he was not content merely to examine it and prove its origins to his own satisfaction. The experimenter in his soul wanted to confirm that paper could be produced from algal mats, and so Leeuwenhoek went to a couple of places where he knew growths of these 'plants which had come from water' could be found. He dried them down in front of the fire to form papery substitutes, and wrote up his account in his letter. To me this was an added benefit, for Leeuwenhoek had made many lengthy observations of pond microbes in the same algal growths, and these organisms were imprisoned in the dried material. All we needed was a means of putting back the realism into these dead, dry remains. For this purpose I allowed minute portions of the dried matter to reconstitute in sterilised lake-water. The specimens were left for 36 hours. At the end of the first day the dried cells were already expanding and regaining the form they had displayed during life, and by the second evening they were succulent, glistening, rounded and remarkably fresh-looking. By manipulating the specimen material on the slide it was possible to tease portions out for closer examination, and in this way I came face to face with some of the microscopic organisms Leeuwenhoek had been first to discover.

Perhaps the most delicate were the diatoms, tiny single-celled algae that build for themselves a supportive skeleton of glass out of the silicates dissolved in the water where they live. As objects of microscopic study, diatoms are without parallel and their ultra-fine structure (which is often beyond the resolution of the best optical microscope, and can only be studied by the electron beam) is regular and meticulous. Diatoms may move along by circulating their protoplasm like an armoured vehicle crawling on caterpillar-tracks, and some can attain a speed – size-for-size – that is equivalent to an Olympic swimmer. They produce oil droplets as energy stores, and were very likely the organisms in the palaeozoic past that laid down the beds of dead remains from which our modern oil-fields originated. Although the diatoms in the Leeuwenhoek specimens are striking and clear, he does not seem to have realised that they were a distinct form of life. That may be because their fine structure is so *very* fine that you would need a refined microscope to make out the details, and Leeuwenhoek was always watching for more exciting and immediately-obvious organisms that crowded into his view. But he knew they were there, right enough, for one of the drawings he sent to London shows a root of a duck-weed plant with typical diatoms clustering around it.

The largest of the organisms I saw were water-fleas, which are readily visible to the naked eye, but which need a microscope to elicit internal structures. These ghostly forms seemed to peer back at one: their eyes, their fertile brood-pouches with half-formed young, their last meal visible through the translucent body wall, all well preserved. Somewhere in the middle of the size scale came the rotifers. These organisms bear one or more circular arrays of tiny beating cilia in life. These hair-like projections whisk back and forth in the water in a strict order, so that the entire structure seems to be a turning gear-wheel.

Leeuwenhoek's description of these entrancing creatures is a masterpiece of vivid documentation. He obtained his rotifers from the same material that he had used in one of the paper-making experiments and, after reconstitution, here they were once more. Rotifers are remarkably resistant to drying – a fact that Leeuwenhoek was the first to document and to test experimentally – and they have been shown to survive for a century or so. The single rotifers that I was able to tease out of the mass of swelling pond-matter looked in good condition, with tiny orange globules inside their bodies (the oil to which I referred earlier) and it may be that some of those might be returned to life. This would establish a longevity record for dried and desiccated animals, and – most important for many of the scientists who have followed the research and cooperated so enthusiastically in it – it would be immensely intriguing to examine some of Leeuwenhoek's own microscopic subjects, returned to life after such a lapse of time.

Apart from deriving information from the specimens in these modern microscopes there were other targets in sight. One of these was perhaps the most satisfying experiment I have ever undertaken. I flew to Holland with two sections each of the cork and the elder-pith, and presented them in front of an astonished University Museum Director at Utrecht, Peter Hans Kylstra. Nothing was still being said publicly about the specimens, as the Royal Society were awaiting the proofs of my paper on the discovery that was to appear in their journal *Notes and Records* and this had to take precedence. All Peter knew was that I had some surprising news for him, and that I wished to request access once again to the little Leeuwenhoek micro-scope that lay in its box in the safe. He found for me an old forensic photomicroscope in a cupboard behind the display stands, and offered it as a means of supporting the Leeuwenhoek microscope. All it needed was an extension which I designed on an envelope and which the department made up for me that afternoon. This gave me a firm base for the little microscope, and a stage on which the specimens could lie. I still needed a support for the brass plate of the instrument, and used a Welsh-language Kidney Donor Card with a hole punched in the end to accommodate the bulge where the lens was situated. With this in place, and the Leeuwenhoek micro-scope neatly in position, I lifted one of the original sections on

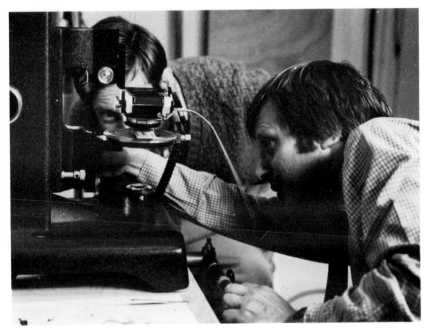

FIG. 9 The first historic demonstration of Leeuwenhoek's sections through his original surviving microscope. The author is here being assisted by Dr Robert Frederik of the Utrecht University in this photograph by Jaap Stolp.

to a small glass slide with a pair of surgical forceps, adjusted the light source that was built in to the base of the stand, and carefully focused the image.

For the first time in three centuries, here was a view through one of Leeuwenhoek's own microscopes, of his personally prepared sections. It was a heady moment, and others in the room gathered round for a glimpse. As rapidly as possible I captured the images on film, and then, changing the specimen, repeated the process. Within an hour I had a roll of film – which Jaap rushed to develop – bearing the images of the kind Leeuwenhoek would have seen. That moment had the feeling of a culmination about it: this was the climax of a frenetic period of activity all conducted with a clandestine atmosphere of excitement about it. Later I spent a little time taking colour photographs of some other specimens through the Leeuwenhoek lens – fungus spores, the tiny spines on their surface showing clearly; fresh blood cells from a finger-prick, showing the nuclear structures in the white cells; a human hair; some living mouth-bacteria . . . and those images too were stored away on film.

The next investigation had occurred to me whilst this was going on: I decided to take the sections which had been photographed with the Leeuwenhoek lens and attempt to find the same cells under a modern microscope, and then in due course under the scanning instrument. This would provide clear evidence of the exact size of detail that Leeuwenhoek could have observed himself, and that was going to solve several puzzles in one since the problem of just what these microscopes *could* reveal had remained current ever since Leeuwenhoek first came into contact with the world of science. Each section posed a technical difficulty, because finding a field of view of half a dozen cells in a specimen composed of many thousands is difficult. When you bear in mind the fact that the cells all look similar, that they can be in any orientation around the full 360° circle (to imagine that, try recognising an upside-down map outline of an unfamiliar country) *and* that the section stood a 50:50 chance of being backwards on the slide by now, so it would have presented as a mirror image, it was clear that a diligent task lay ahead. With the cork section it was not too complicated, for I had chosen cells in the thinnest region of the section and that was reasonably easy to find. The elder pith was a more difficult matter, but the characteristic shapes of the cells I had first seen through Leeuwenhoek's microscope could be recognised and I was able to adjust them to my satisfaction for micrographic study. You tend to get to know your way round a microscopic specimen, like finding your way in a wood; and the entire problem was resolved within a few minutes.

Later I took the same sections from the glass slide, fixed them on to the aluminium stub used in the scanning microscope and prepared them to be coated with gold. Finally we saw the same cells on the screen of the electron microscope, and took exposures at different magnifications from which it was possible to measure the finer details that the Leeuwenhoek lens had disclosed. The results were most impressive – the lens, with a magnification that has been measured at ×266, could reveal fine fungus hyphae that were less than a thousandth of a millimetre wide. It was a superb test of an old lens, and showed

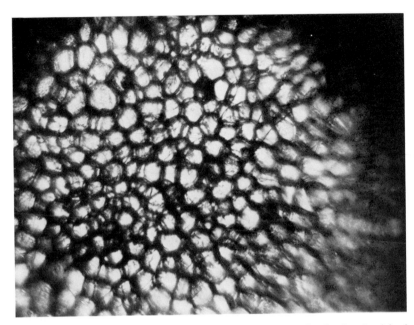

FIG. 10   The image of Leeuwenhoek's section of cork obtained with the original microscope at Utrecht. This study, magnified ×266, compares surprisingly well with modern microscopic images. The finest structures visible are less than 1μm (1/1000 millimetre) in thickness.

in practice how successful Leeuwenhoek – the master lens-maker – truly was.

After the initial investigations were completed there arose the question of putting the specimens into their historical context. What had Leeuwenhoek himself said about them, for instance, and what was known of their history since they had first arrived in London? The most recent reference to the specimens was to the first packet I found. It came in a paper published by Professor F. J. Cole in 1937, who was writing about Leeuwenhoek's 'rough hand sections' in a paper published in the *Annals of Science*. His paper was published in two parts, altogether nearly one hundred pages long, and buried in it he wrote: 'some of the sections [Leeuwenhoek] cut with a sharp shaving razor still survive'. The same packet was mentioned by Dobell, in a section of his biography on Leeuwenhoek's methods of measuring microscopical dimensions. There he mentioned a 'little packet affixed to an early letter' and noted that it was 'intact to the present day'. Neither of these important and influential men made any reference to the other six packets, however.

The existence of the three dried algal mats with their enclosed examples of Leeuwenhoek's rotifers and other microbial forms would have been of the greatest interest to Dobell, whose book concentrates on Leeuwenhoek as microbiologist, and I regret that his own investigations into the dawn of microbiology missed this opportunity to examine some of the original material, just as they omitted an examination of Leeuwenhoek's lenses. These algal specimens, however, were the only ones to be noticed by the Dutch translators of Leeuwenhoek's letters. The fact that there were three small envelopes containing dried algal material was noted, though no biologist was brought in to analyse the specimens.

In the case of the cotton-seed specimens, the matter edges towards farce. Leeuwenhoek wrote in his letter of 2 April 1686 that he was sending some specimens to the Society, and the translator of the *Collected Letters* noted ruefully that this 'slide' was no longer in the possession of the Royal Society. The mention of a missing 'slide' is clearly an elementary blunder, for slides are more modern affairs and no one would imagine that Leeuwenhoek would have sent such a thing to London back in 1686! It is also surprising that the same question had not been raised in connection with the letter of 1 June 1674, which accompanied the sections, for that letter too said Leeuwenhoek was sending over some samples of his work, but the *Collected Letters* have nothing to say about that.

In fact, I found that the specimens *were* with the letter all the while. They were pasted near the end of the final page, alongside the writer's signature. A footnote to this passage inserted by the translator notes that two 'drawn rectangles' appear alongside the signature and adds 'in one of them, the words "nine seeds of a cotton tree . . . ." have been written.' The reason the packets looked like rectangles to the translator is that he was working from microfilm copies, where the packets might indeed have looked like something 'drawn'. But, having noted that the specimens were missing from the Royal Society, and being faced just a few paragraphs later with rectangles captioned with perfect descriptions of what was missing . . . it seems incredible that nobody bothered to check.

Work on these specimens will continue and they still hold

secrets of their own. There is one final word that should here be recorded, though, and it is only right that it should be from Leeuwenhoek himself. This is how he described the sections he was sending to London:

> Which kind of progress of growing I apprehend may in some manner be seen in the pith of *Wood*, in *Cork*, the pith of the Elder, as also in the white of a *Quill*, of which three last I have sent you and your curious Friends some small particles, cut off with a sharp razor.

A strange choice of objects? So you might think. Let us turn to Hooke's *Micrographia*, the book which I have long believed served as an inspiration to Leeuwenhoek at the dawn of his illustrious career. On p. 133 appear the following words:

> Nor is this kind of Texture peculiar to Cork onely; for upon examination with my *Microscope*, I have found that the pith of an Elder, or almost any other Tree . . . have much the same kind of *Schematism*, as I have lately shown that of Cork . . . The pith that also fills that part of the stalk of a feather that is above the Quil has much such a kind of texture.

The same specimens, and listed in precisely the same order. Who can doubt that Leeuwenhoek was no self-taught visionary with a considerable 'silent period', but a convert to the cause, inspired by Robert Hooke. The fact that Leeuwenhoek spoke no English is neither here nor there: his letters often reveal that he knew people who provided translations on request. The unexpected find of his own microscopical preparations provided an additional bonus – final testimony in support of my view that Hooke was the genius from whom Leeuwenhoek's life's work stemmed.

How surprising it is to realise that, though Leeuwenhoek gave to science and humanity a solid half-century of devotion, he did not begin to communicate his work until he was already a middle-aged man. Leeuwenhoek was over 40 when his first primitive experiments were brought to the attention of the London establishment, an age when many modern scientists are starting to think about resting on their laurels. It was a late start, but he made up for lost time.

4 | Missing Microscopes

It took a considerable time for Leeuwenhoek's use of the simple microscope to gain widespread recognition, and that only came about because of the astonishing versatility and accuracy of his research work. The collected editions of the letters which Leeuwenhoek himself saw published began with the letter No. 28, dated 25 April 1679, and these were the first of his descriptions to have a wide audience. Of the previous 27 letters, only two were published in full at the time, and a further 11 appeared in shortened form, usually in the Royal Society's *Philosophical Transactions*. Yet by September 1674 he was already publishing observations that showed him to be a superior microscopist to Robert Hooke. It was in this month that Leeuwenhoek announced the discovery of microbes in pond water, and his clarity of description makes it possible to know which organisms he was observing even though his terms were culled from everyday language, since no one had ever seen these organisms before.

It would be wrong to imagine that Leeuwenhoek was the first person in history to record the fact that a lens could reveal small living organisms. Though there is no way of knowing when the idea first arose, the earliest published account of which I know dates from 1508, when Alexander Benedictus (who, like Fracastoro, lived in Verona) wrote of tiny *lumbrici* or 'wormlets' being found in the skin, kidneys, lungs, teeth and also in cheese. He gives the impression of having been perfectly familiar with the notion that minute living organisms could spread disease and he seems to have known that a tiny mite living in the skin caused scabies. Leeuwenhoek's mastery of disclosure of this invisible universe gave science the lead, and whereas earlier writers were often guessing or were led by superstition, Leeuwenhoek's objective approach certainly warrants his mantle as the 'father of microbiology'.

What of Robert Hooke, whose inspiring observations in *Micrographia* had been published years before, in 1665? Were not his activities comparable, and even greater? I think not, for though they amount to a magnificent body of work, he was not in Leeuwenhoek's league and his observations with a compound microscope fell short of the results that the simple microscope could provide in several important respects. Whereas Leeuwenhoek was investigating truly microscopic phenomena and was hunting for previously unknown forms of life, Hooke was magnifying objects that were already familiar. His aim was the production of enlarged and dramatic views of the conventional, rather than searching out the unknown. Thus he produced eye-catching pictures of a louse, a flea, a stinging-nettle leaf, the edge of a razor, a human hair; Hooke's interest was in making a fascinating book which everyone would want to buy, and in burying in it some observations in other areas which he would not have readily published for a wide audience any other way (his views on gravity, refraction, the nature of light and many other topics are all included in the text of *Micrographia*, which even has some fine studies of the heavens, showing stars and the lunar craters, for good measure).

Leeuwenhoek was almost doing the opposite: he was not so interested in the familiar, but in discovering entirely new things; and his aim was not to impress large numbers of people – indeed, he turned away some eminent visitors who called in due course, unannounced, at his house – but to communicate with other scientists who might be interested. Whereas Hooke took pains to find out that there was a ready market for his ideas, Leeuwenhoek made it plain that he did not mind in the least if others disputed his claims and emphasised that his only aim was personal satisfaction.

The great divide between the work of the two men centres on the magnifications they used. Hooke's compound microscope gave him a direct magnification of only twenty or perhaps fifty times at the most, and for much of his work he was using a degree of enlargement that today would be regarded as the 'low power' of a conventional laboratory microscope. He was limited in this respect because lenses in train tend to magnify the optical errors in each other, as much as they magnify the image; and it was not until the corrected lenses of the Victorian era became popular that this could be improved upon. Leeuwenhoek was working with magnifications that a modern research worker would find perfectly adequate for most normal purposes. His lenses were up to ten times as powerful as Hooke's compound system. Let nobody doubt Hooke's great importance as the populariser of microscopy, but neither should we fail to acknowledge that it was from this time on for the better part of two centuries that the high-power simple microscope, the Cinderella instrument of which so few people have heard, was the key to progress.

The nine surviving microscopes are fascinating little objects. When I went to examine them in 1980 I soon discovered that the lenses were already being investigated by a Dutch lens specialist, Dr J. van Zuylen, in a spare-time programme of research. Since that time we have pooled our ideas on the lenses that Leeuwenhoek made, and his special knowledge of lenses has thrown up findings published in 1981 to which I direct enthusiasts who wish to know more. Several people have measured the magnification of the lenses, and the figures vary depending on the circumstances of the experiment. I have measured some of them myself, and obtained results that are slightly at odds with van Zuylen's figures. If van Zuylen's results say (as they do) that the magnification of the Utrecht lens is ×266 then that is the figure I accept. We are in either event only speaking of detail, not orders of magnitude, and it is the capacity of the lens to see fine detail – its resolution in practice – that matters above all, so one figure is in practical terms as good as another.

Of the nine microscopes, five prove to have a magnification of more than ×100, and not one was less than ×69. One of the microscopes which belongs to the Boerhaave Museum in Leiden, but is now to be found in a glass case set into the wall of the Laboratory of Microbiology in Delft, has long since lost its lens. In each case it was assumed that the lenses were made from selected fragments of glass that were then ground and polished, and though this proved to be so for seven of the lenses, for one it was not. This lens – the example at Utrecht and the most powerful of them all – seemed to have been made by a process of melting and showed no signs of being ground. There is a documentary record for this possibility, for when von Uffenbach called to visit Leeuwenhoek in 1710 he quoted him as saying 'that he had succeeded, after ten years speculation, in learning a useful way of blowing lenses which were not round.' Uffenbach dismisses this as impossible. He wrote: 'It is not possible to form by blowing anything but a sphere or a rounded extremity.' However, when van Zuylen examined the lens he found that it was indeed not 'spherical', the surface was a complex curve more like a parabola and this gives a wider clear field of view than you would expect of a ground, spheroidal lens. One answer to how it was made was put forward by van Zuylen, when he and a friend, Mr J. Nieuwland of Delft, softened a length of 15 mm glass tubing and blew a bulb in the middle of it. Then they melted away one of the projecting ends of the tube in the glass-blowing flame, until the bulb was as round as an electric lamp but with a small glass 'pip' at the end, where the tube had originally been. The bulb could then be broken and the 'pip' extracted. This, they found, was remarkably similar to the lens that Leeuwenhoek had fitted to the Utrecht microscope, and it gave satisfactory results when used to magnify a microscope slide. The result of this alternative approach provides a perfectly workable solution to Leeuwenhoek's desire to blow non-spherical lenses, and the Utrecht lens indicates how well his results worked in practice.

What about the microscopes themselves? They are all based on similar principles of construction, the lens being sandwiched between two metallic plates which are riveted together at the corners. A long screw passing up from the bottom of the microscope supports a stage block, into the front of which is set a focusing screw. This projects through the block and pushes against the microscope plate when screwed home, moving the stage block slowly away from the lens. By turning this screw carefully, the object may be brought into focus. On the top of the stage block is a third screw terminating in a blunt point. That is where the specimen was set, either by being impaled, or stuck in position with glue. Into the side of this fits a little handle which enables the specimen-holder to be turned around, so that the object can be examined from all sides.

An enquiring mind will wish to look further than this. Three possible questions at once arise:

(1) Is it possible that any of the nine microscopes are not authentic?
(2) Are there any other lenses not in the list, or any other Leeuwenhoek microscopes not included among the nine?
(3) Might there be other microscopes awaiting rediscovery?

In each case the answer is 'yes'. Authenticity is not an easy matter in documenting the origins of historic artefacts. There is the maker's name – except that it can be forged. Then there is an individual style – which can always be imitated; or a method of construction, which can always be copied. Records become lost, notes mislaid or even falsified, memories are faulty, and written accounts can be unintentionally misleading because the writer assumes knowledge about one aspect of which a researcher, many years later, knows nothing.

The provenance of many of the microscopes is reasonably well established. A few can be traced back to the auction of Leeuwenhoek's instruments which was held after he and Maria were both dead. Some have remained in the Haaxman family for generations, purchased at the auction and handed on from there. But there may well be interlopers. One about which I am none too sure is in the collection of the Royal Zoological Society of Antwerp. It somehow does not truly have the feel of a Leeuwenhoek microscope. Artefacts acquire a certain handwriting, as it were, a style given to them by the way the maker worked and – although it is perfectly possible to try to imitate the style – if an object does *not* have that characteristic quality then it makes one wonder if it can be genuine.

The first time a description of this microscope appeared in print was 1945, when Dr Ed. Frison wrote about its lens. The instrument was described as a brass microscope from the collection of Henri van Heurck, a businessman who devoted himself to the collection of old microscopical memorabilia. At the time, Frison wrote of the microscope as being unquestionably genuine and yet nothing was published about the reasons .behind this assumption. The van Heurck archives have nothing to reveal, as most of those were destroyed by fire in the 1939–45 war in Europe, and the first time the microscope appeared in his possession – as far as the published record is concerned – was in the 1914 catalogue of all the van Heurck microscopes, when the collection was being valued for auction following their owner's death in 1909.

In 1891 van Heurck did not possess the microscope. In that year the fourth edition of his book *Le Microscope* was published, and in the new historical introduction he used as an illustration of a Leeuwenhoek instrument some woodcuts modelled on the famous diagram of the Utrecht microscope that had been drawn by John Mayall around 1885. If he had owned a genuine Leeuwenhoek microscope he would certainly have used a study of that as his illustration instead. By coincidence his collection of historic microscopes was put on exhibition at the great Exposition de Microscopie held in Antwerp that year. Both the Haaxman microscopes, and the example from Utrecht, were on show, and it is clear that if one had been in his possession it would also have been on display. At some time between 1891 and 1909, therefore, this instrument entered his collection. A rediscovered Leeuwenhoek microscope in Belgium would have been a considerable high-point in his activities, yet there is nothing to suggest that this exciting episode ever took place. It could be genuine; but the lack of any provenance and the fact that the microscope has several features that make it different from the various designs Leeuwenhoek used to produce, together tend to make one question its authenticity.

There are certainly other lenses that are not on van Zuylen's list. In the collection of the Boerhaave Museum in Leiden is a small red morocco leather case which contains five small lenses. They are mounted in brass slips and they range in magnification from ×30 to ×167. There is a microscope in the collection which they seem to fit. It is similar to a design by Leeuwenhoek, in which a simple microscope is mounted on to a holder for a glass tube in which a fish, usually an eel, could be studied and its blood circulation examined closely. This form of instrument was called an *aalkijker*, or eel-watcher. In P. J. Haaxman's biography published in 1875 (pp. 34 *et seq.*) he described how an instrument like this had been presented to Peter the Great, Czar of Russia, when the two men met in Holland. Haaxman relates that the *aalkijker* had been brought back from Russia by Johan de Gorter (1686–1792) or his son David (1717–1783) who lived for seven years in St Petersburg. Could this be the same instrument? There are clear reasons why the suspicion should be raised, but against them stand the fact that the lenses are not of high quality (so that they would not seem to be the kind of gift that Leeuwenhoek would wish to present to such an illustrious visitor), and that the microscope itself – though similar to Leeuwenhoek's designs – may not have been made by him. It is similar to one made by van Musschenbroek and could have been a prototype of the microscopes he later produced. Some of those survive, bearing his own mark, and they are similar to this example.

The question of microscopes still to be rediscovered is a profoundly interesting one. There is a chance that some may still be hidden in corners. There could yet be an entire cabinet of Leeuwenhoek microscopes, for one set went missing in London a century or so after being posthumously bequeathed by Leeuwenhoek to the Royal Society. There are records of others being presented to members of the British Royal Family which it should be feasible to trace. One is encouraged in this view by the fact that one silver microscope was discovered a few years ago in a box of odds and ends in a Dutch laboratory. It seems that the Zoological Department of Amsterdam University moved from one building to another in 1924 and a box of discarded glassware from the laboratory was put on one side. Among the items was an original silver microscope, identifiable by a hallmark stamped onto the microscopes when this was a legal requirement between 1814 and 1834, and also by the fact that the focusing screws were found to be interchangeable with those on the microscope's twin in the Leiden collection. The threads Leeuwenhoek made were of his own distinctive pattern, and the fact that the screw on the rediscovered instrument fitted into its counterpart at Leiden showed they were both made by the same maker in his own individual die. With this precedent in mind it is certainly possible that other Leeuwenhoek microscopes might yet come to light.

Most of them are very likely in Holland, if they have not been discarded over the years, or used as toys in past centuries and thrown away. It has been calculated that Leeuwenhoek must have made some 500 or more microscopes, so it is possible some may have survived. Others are possibly in London. A pair were given to Queen Mary (of William and Mary fame) to commemorate her visit to Delft to meet Leeuwenhoek. The royal microscope collection was passed on to the Science Museum for custody and it does not seem they exist anywhere there – though it might be that an uncatalogued box of oddments still does contain them. Sir Robin Mackworth-Young, the Librarian to the Queen at Windsor Castle, informs me that the only optical instrument in their collections is a binocular microscope presented to the Duke of Edinburgh by the Royal Microscopical Society when he was the Society's president, and Geoffrey de Bellaigue, Surveyor of the Queen's Works of Art, tells me that they are not in their collections either.

But of greater interest are the missing microscopes that Leeuwenhoek left to the Royal Society and which were sent over to London after he died. Martin Folkes, then the Royal Society's Vice-President (he was President 1741–1753) wrote a description of the microscopes after they were received safely in London:

> The Legacy consists of a small *Indian* Cabinet, in the Drawers of which are 13 little Boxes or Cases, each containing two Microscopes, handsomely fitted up in Silver, all which, not only the Glasses, but also the *Apparatus* for managing of them, were made with the late *Mr. Leeuwenhoek's* own Hands: Besides which, they seem to have been put in Order in the Cabinet by himself, as he design'd them to be presented to the Royal Society, each Microscope having had an Object placed before it, and the Whole being accompany'd by a Register of the same, in his own Hand-Writing.

There are some earlier published remarks that surely apply to the same collection of microscopes, as when Uffenbach wrote in 1710: 'Mr Leeuwenhoek afterwards fetched some cases, in each of which were two *microscopia* . . . of quite a simple structure.' Leeuwenhoek's own description of the legacy came in his letter of 2 August 1701, in which he says:

I have a very little Cabinet, lacquered black and gilded, that comprehendeth with in five little drawers, wherein lie inclosed 13 long and square little tin cases, which I have covered with black leather; and in each of these little cases lie two magnifying-glasses (making 26 in all), everyone of them ground by myself, and mounted in silver, and furthermore set in silver, almost all of them in silver that I extracted from the ore, and separated from the gold with which it was charged; and therewithal is written down what object standeth before each little glass. This little Cabinet with the said magnifying-glasses, as I may yet have some use for it, I have committed to my only daughter, bidding her send it to You after my death, in acknowledgement of my gratitude for the honour I have enjoyed and received from Your Excellencies.

The startling fact that these microscopes had been lost from the Society's care came to light officially on 5 April 1855, when Sir James South wrote from the Kensington Observatory to complain that he had asked to see the microscopes of Leeuwenhoek, but had been told no one knew where they were. The Society's Council reacted by minuting their decision to appoint a committee to investigate the disappearance 'consisting of the Officers, with Mr Tite, Mr Wheatstone, Mr de la Rue, Mr Bell, Mr Miers, and Mr Grove' to enquire and report on the state of *the Society's manuscripts*. The fact that there was an immediate emphasis on missing documents, rather than the missing microscopes, is the first disappointment. A second one is the revelation that South clearly knew more than it appeared. In June the Secretary wrote to say that he understood 'the microscopes bequeathed to the Royal Society were still in existence and that you knew in whose possession they were . . . so would you favour me with any information you feel at liberty to communicate.' South was having none of this – he wrote back tersely and said: 'I at present decline correspondence, except with the President and Council, on any of these interesting objects.' To mollify him, Council resolved that the Secretary should write once more, but 'in the name of the President and Council', to ask for information leading to the recovery of the instruments. On 28 July 1855 South replied in more detail:

I have the honour to state, that many years ago, wishing to Examine the Leeuwenhoek microscopes, I was officially informed by Mr Stephen Lee . . . that they were lent to Sir Everard Home, and this, I grieve to say, is all the information I possess.

South did write once again on this subject, asking to be informed of any findings, and the minutes of Council show that they: 'Read a letter from Sir James South on the subject of the Leeuwenhoek microscopes,' but nothing was ever found. The microscopes remain missing to this day.

The whole subject is shrouded in mystery. Most tantalising of all is a portion of the letter of 28 July 1855, where South explained that he had even once been to the house of the 'Gentleman alluded to' (no doubt Sir Everard Home).

He received me most courteously, showing me most unreservedly all his antique scientific curiosities, amongst them the microscopes which he had long greatly prized as having been the microscopes of Leeuwenhoek, but which I convinced him they had never been, by placing in his hands the sketch and model of those of Leeuwenhoek which I carry in my pocket-book.

The drawing he carried round with him was taken from Henry Baker's book *Employment for the Microscope*, which was published in 1753 and contained the first-ever published drawing of a Leeuwenhoek microscope.

The problem here is that the drawing Baker published does not look much like a typical Leeuwenhoek microscope, at least according to the idea of a 'typical' microscope which we have adopted since Leeuwenhoek's death. This fact – added to the confusion over the exact dates when the cabinet of the little instruments went missing, the failure of Sir James South to identify exactly who it was he had approached, and the apparently unresolved 'investigation' carried out by the Committee who seem never to have produced a report – leaves the question of their fate very much in the air.

Or does it? I think we might, after all, be able to gain an idea of how the microscopes looked. It is likely that the missing microscopes were significantly different from those with which we are familiar. And there is at least one further possibility that might be considered before they are given up as irretrievably lost. The difficulty is that our conventional view of a Leeuwenhoek microscope has been so heavily biased by the well-known Utrecht instrument. This was the one of which the first detailed engravings were made, and they have been reproduced so often in one form or another that it is tacitly assumed that this is the only kind of microscope Leeuwenhoek made. In fact he produced many different instruments, even if they were loosely based on the same general idea. Some of them had more than one lens, indeed the famous portrait of Leeuwenhoek showing him with a microscope in his hand suggests that he made an instrument with three lenses side by side, set into a rectangular body considerably larger than the 'typical' type. The Utrecht microscope, I believe, represents a late version of his design career. Its focusing screws are sophisticated and the body has a solidity lacking in some of the others. The main supporting bracket is fixed off-centre, which means it can rise vertically to the stage (if it is mounted centrally at the base, then it has to be tilted sideways to allow the stage pin to stand centrally). If Leeuwenhoek was initially stimulated by Hooke's *Micrographia*, we can clearly speculate that his earliest-ever microscopes would have been modelled on the specifications Hooke gave in his Preface: namely, a polished lens set into a perforation in a metal plate. If that is the case – and it is the view I advance – then his microscopes must have shown a gradual evolution and it would be possible to fit the existing microscopes into a scale that related their construction to the passing of time. On this basis, a centrally fixed main bracket would be earlier than one set off-centre, but vertical. Similarly, since silver is so much easier to work than brass, silver microscopes might logically precede brass ones. High-magnification lenses would probably be made later in his career than low-magnification examples, and so forth.

These are not absolute criteria. There is no reason why Leeuwenhoek should not have made relatively low-magnification lenses late in his career. But that is not the point. The argument is that he would have been most unlikely to make well-finished high-magnification lenses *early* in his career, and that is the key. Similarly, a well-finished brass microscope would be more likely to fit late in the time-scale, rather than early.

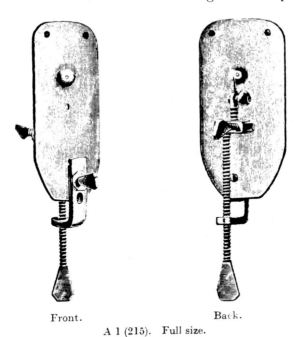

Front.   Back.

A 1 (215). Full size.

Fig. 18.   Fig. 19.

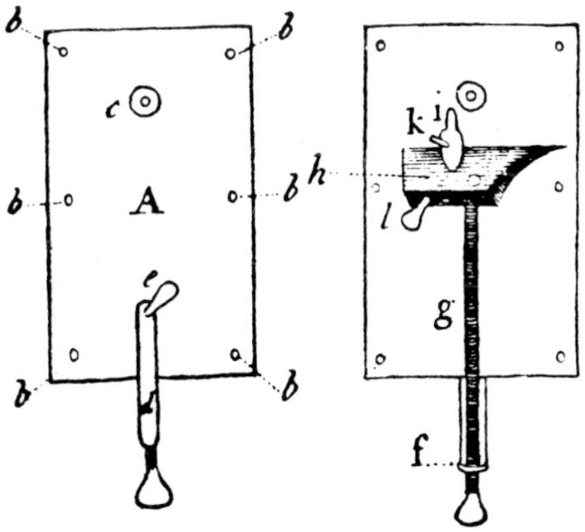

A 1 (215): Hoole's figures.

FIG. 11   The famous Mayall engravings of the Utrecht microscope have provided the conventional view of a Leeuwenhoek microscope for a century. Baker's figures (in this plate, taken from a 1928 publication, they are wrongly attributed to Hoole, who merely reproduced them) are the only surviving studies of the Leeuwenhoek microscopes sent to London by his daughter.

There are other clues from contemporaries, when they mentioned that Leeuwenhoek had recently succeeded in making a blown lens, for instance; or when they record that he was actually making silver instruments. Between the primary sources, the contemporary accounts, and the surviving microscopes, we could begin to string the types into some kind of chronological sequence. This makes it even more important to know what the missing 26 instruments looked like. The only drawing of them was the odd little sketch diagrams that Baker published, and because of the strange perspective people have dismissed them as of little value. Dobell declined to include them, saying there were more 'instructive data now available'. Van Zuylen dismisses them, saying they were probably drawn from memory and were not necessarily accurate. What does Baker himself have to say?

His first detailed account dates from 1743 when he described the microscopes sent to the Royal Society and adds, in a footnote, 'At the Time I am writing this, the Cabinet of Microscopes left by that famous Man, at his Death, to the *Royal Society*, as a Legacy, is standing upon my Table,' as evocative a statement as you would find in all the literature on the missing microscopes! The diagram, with its peculiar way of drawing, was published in 1753 and, although it has been reproduced many times, less well known are the words that Baker wrote about the microscopes at the time. I think they hold the clue to the appearance of the lost collection, for they disprove once and for all that he was drawing purely from memory, or guided by guesswork:

> The Curious will be pleased to see a Drawing of them, *taken with great Exactness* from those in the Repository of the Royal Society, which are all alike in Form, and *differ very little in Size from this Drawing*, or from one another. (My italics)

I looked again more closely at the diagram, taking them not merely as sketches from memory but as accurate scale drawings which, even if their perspective is idiosyncratic in the manner of many drawings of the time, give us the details of the dimensions and the construction of the missing microscope collection.

The fact that they are of silver and have a centrally

mounted main bracket suggests that they are relatively early instruments. The design shows a distinctive feature in the microscope stage, shown as triangular in section and unlike the conventional Leeuwenhoek microscopes featured in the standard texts. Such a stage does, in fact, exist. The silver instrument now in the collection of the Deutsches Museum in Munich has a stage that is triangular in cross-section. The stage pin has a peculiar design, but we can find a similar example on the microscope in the private collection of Dr J. J. Willemse of Rotterdam. The shape of the main screw and its

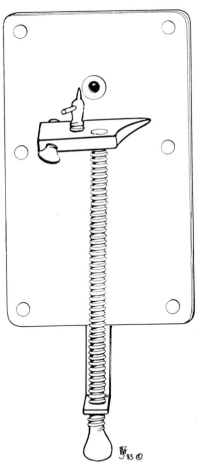

FIG. 12 Using the date given by Baker and described in the text, the appearance of the missing Leeuwenhoek microscopes may be recreated. It is possible that they still exist in a neglected cupboard in London; the last known possessor was Sir Everard Home, a surgeon, around 1820.

rounded, flattened handle is unlike the conventional view, but similar to that of the microscope at Munich; and the Munich instrument also shares a similar shape and an identical riveting pattern where the two plates of silver are fixed together.

If the Baker diagrams are accurate, as their artist insists, then we can obtain from them a number of details of the missing microscopes. They were made of silver plates approximately 36 mm × 22.5 mm (closest to those dimensions is the Rotterdam microscope, measuring 39 mm × 22 mm). The pitch of the long screw – i.e. the length of each turn in the thread – is the length of the screw, 31 mm, divided by the number of turns, 44, equalling 0.72 mm. This fits well with the other types known, the average pitch of the nine microscopes being 0.75 mm. As for the cabinet, it could have been of any one of thousands of designs. Leeuwenhoek wrote of it as an 'Indian cabinet' though this was a misnomer. The cabinet was of the popular type decorated with what we would now call Chinese lacquer and imported from the Dutch East Indies. Many such cabinets have survived in museums, and there are numerous examples in households to this day. I have one myself, passed on from a great-grandparent.

Using this information, it is possible to reconstruct the appearance of the missing microscopes. There is evidence to support the fact that the last man believed to have them – Sir Everard Home – was greatly interested in such single-lensed microscopes, for in 1823 George Bentham returned to England and met many botanists and microscopists of the time and wrote that Dr Everard Home was greatly interested in the performance of such instruments. That is around the time the microscopes were last seen by the Royal Society.

Home's record in other respects suggests that he certainly *could* have been unscrupulous enough as to keep the microscopes. He is most renowned as the student of that innovative and pioneering surgeon, John Hunter, whose assistant he became in 1773. After Hunter's death, the mass of papers and descriptive work amassed by him over a life-time of industry was bequeathed to the College of Surgeons in 1800. Home arranged that William Clift – curator to the college – should send them instead to his own home, where he promised to index the papers. The college asked for the work to be completed on many occasions, but they were repeatedly fobbed off with the excuse that they were required for a little while longer. A short synopsis of the index was published in 1818, but it soon became apparent that Home's real reason for keeping the notes was that they had become his own source material for the preparation of papers in which Hunter's discoveries were presented as Home's own work. It was an unscrupulous and shameful episode. In the end, when the pressure to examine Hunter's papers grew too great, Home burned them at his home. When he broke the news to Clift, he apparently said he had come close to incinerating the entire building in the process.

This immediately suggests one possible, tragic fate for the missing Leeuwenhoek microscopes – perhaps they too were destroyed in the conflagration. The matter would have been common knowledge amongst the circles in which he moved, and the loss would have been accepted as an unfortunate accident. Since Home used to insist that Hunter had specified that all his papers should eventually be burnt, he could have accounted for the microscopes at the same time.

There is also the interesting fact that Home did not personally carry out most of the microscopy which appears as illustrations in his works. Instead, he had technical workers who carried out this labour on his behalf. His anatomical drawings were undertaken by Clift, whilst the microscopical studies were the work of Fritz Bauer. The standard of work of this admirable man is of the highest order, and this might have been aided by the excellent image generated by single lenses, rather than compound systems. That would tie in well with Home's interest in the microscope cabinet that Leeuwenhoek had sent to London. It is possible that Bauer would have used the microscopes for his detailed studies and they may have spent more time in his hands than those of his employer. 1823 is the year which saw the publication of Home's work involving Bauer's drawings at high resolution (itself implying the use of simple microscopes), when Bentham recorded Home's interest in Leeuwenhoek-type microscopes, and about the time that the instruments were taken from the Royal Society. Although they could never reveal as much new knowledge about Leeuwenhoek's work as his marvellous specimens have done, their rediscovery would be an immensely satisfying and exciting event.

5 | Birth of a Science

We have seen how Antony van Leeuwenhoek became a master at microscope manufacture, or at least lens-making, and how he managed to cut fine sections, to carry out experiments, and to make observations that were of the greatest accuracy. As I have said, a great proportion of his original writings have yet to be published in translation, so that no book like this could attempt to tell the whole story of his work. But, since the world of modern biology rests so surely on his pioneering efforts, it is time that we began to appreciate the extent of his labours more fully. The popular idea that he was a lightweight and nothing more is unjust. But, since it may have come about through the widespread ignorance of Leeuwenhoek's work, perhaps an outline of some of it will help to set the record straight.

The science of microbiology was born in August 1674, when Leeuwenhoek was travelling in a boat across a small lake two hours from his home. 'The lake is known as Berkelse Mere,' he wrote at the time, 'and in many places its bottom is marshy or boggy.' What drew his attention to the water in the lake was a strange cloudiness that developed every year during the late summer, and which the local inhabitants put down to the effect of heavy dew. It has already become plain that

Leeuwenhoek was a realist who was not given to accepting fanciful, metaphysical explanations for rational phenomena. His inspection of the sporangia of the mildew fungus was inspired by Robert Hooke's insistence that no 'seeds' could be found and his logical conclusion that fungi could arise spontaneously. Leeuwenhoek was sure there must be 'seeds' for any plant to propagate and he simply set out to find them. It was the same at Berkelse Mere. Leeuwenhoek was not impressed by the folklore explanations and so he collected some of the water in a glass bottle and took it home with him. Next day he looked closely at the water with one of his microscopes, and his description of what he saw is so beautifully clear, so entrancingly captivating, that it is in most cases possible to know in an instant which organisms he was looking at that day. Here are his words, translated into present-day English:

> Floating in the water were various earthy particles, and some green lines coiled into spirals and evenly spaced, just like the copper or tin coils which distillers use to cool liquids when they distil over. The whole circumference of these lines was about the thickness of a hair from your head. Other bodies showed only the beginning of this structure, but all of them consisted of very small green globules joined together: and there were many small, rounded green bodies as well. Among these there were very many little animalcules* of which some were rounded, whilst others a bit bigger were shaped like an oval. On these latter examples I could see two tiny limbs near the head, and two little fins at the rear end of the body. Other examples were longer than an oval, and these were slow moving and few in number. These animalcules had different colours, some being whitish and transparent, others with green and very glittering little scales; yet others were green in the middle and white at both ends, and some were grey, like ash. And the motion of most of these tiny creatures in the water was so fast, and so random, upwards, downwards and round in all directions, that it was truly wonderful to see. I estimate that some of these little organisms were over a thousand times smaller than the

* The name 'animalcule' – meaning 'minute animal organism' – I have retained here as there is no better modern equivalent word.

smallest mites I have ever seen on such things as the rind of cheese, wheat flour, mould and the like.

I have a private belief – private, though aired in public – that nobody should ever leave school until they have repeated this experiment for themselves. The glistening translucence of pond micro-organisms is one of the most aesthetically pleasing images science can offer mankind, and it is one of the most illuminating and educational single experiences anybody can undergo. Few adults have any real idea of what 'a living cell' really *is*, and this glimpse of the microbial world would restore to all of us a sense of wonder and appreciation of how these normally invisible forms of life go about their lives. We rely on microbes for so many facets of our life: they provide the oxygen we breathe, feed the world's plants, produce foods from bread and cheese to beer and wine, and their dead bodies in chalk and limestone have even shaped the very landscape on which we live out our own lives. I believe they have probably produced the oil-fields and the vast reserves of iron ores on which modern civilisation depends, so it is astonishing that only a small proportion of today's adults have ever seen a living microbe!

Leeuwenhoek was clearly impressed by what he had seen, for this letter – dated 7 September 1674 – was followed by a long and detailed description covering 17½ large folio pages which he sent to London on 9 October 1676. This letter is written in a small and neat hand which is not Leeuwenhoek's own (he must have had an amanuensis copy his notes for him), though he has made occasional corrections to the writing, and the letter is signed by him. His descriptions cover observations made over the previous year or two, and include such vivid passages as this:

When these animalcules began to move, they sometimes stuck out two little horns which were moved around continuously, rather like a horse's ears. The part between these little horns was flat, their body otherwise being rounded, except that it runs to a point at the rear; at which pointed end each had a tail four times as long as the body and looking as thick as a spider's web through my micro-scope. Their tails could be coiled up in a spiral, like a piece of copper or iron wire that has been wound tightly around a piece of dowelling and then taken off, retaining its windings.

This description is perfectly clear, and shows us that the organism he was studying is the bell-shaped *Vorticella*, which lives on the end of a curious spiral stalk which it can contract into a tightly coiled spring if it needs to retreat from danger. *Vorticella* feeds by whisking a current of water through its 'mouth' region and taking in the particles it selects for food whilst rejecting the remainder. Occasionally the organism contracts its spiral stalk and then slowly opens out in a different direction, as though searching a new area for food. All this Leeuwenhoek recorded over the years.

His description gives us another helpful indication too – this time concerning the microscope, rather than the specimen. The fact that he saw the rows of beating cilia around the open bell of the organism like 'horse's ears', and that the stalk of the *Vorticella* was as fine as the thread of a spider's web, shows that he was using a relatively low-powered microscope.

Hooke read this letter with interest and when Leeuwenhoek went on to describe the bacteria he had observed in infusions of pepper, cloves and nutmeg he repeated the experiments in London just to confirm that the Dutchman's observations were reliable. By this time Robert Hooke (having established his precedent with his book on microscopical objects) had turned away to other fields of investigation. To carry out his reiteration of Leeuwenhoek's work, he had to 'fit up some microscopes which had lain a long while neglected, I having been by other urgent occupations diverted from making further enquiries with that Instrument.'

It would be hard to find any serious challenger to Leeuwenhoek, in terms of variety and depth of his interests. For example, the fact that the vinegar eel-worm produces live young, rather than laying separate eggs, was regarded as an important discovery when it was announced by Sherwood in 1746 – but Leeuwenhoek first noted that occurrence in 1676. He also described the different posture adopted by the anopheline mosquito (which transmits malaria, we now know) and its similar relative *Culex*, the gnat. This distinction became of the greatest importance during the attempts to eradicate malaria earlier this century, and it is one that stems directly from

Leeuwenhoek's own painstaking work. He showed that the graceful fresh-water rotifers could apparently survive being dried, and set up some experiments to desiccate them artificially and then record their miraculous return to active life. His studies on the spermatozoa of many species gave rise to many accurate descriptions, and his discovery of countless types of microbes meant that he was the first true microbiologist in history . . . and all of this, together with many descriptions of the circumstances in which he worked, are given in Dobell's biography.

' In fairness, there are some deficiencies in that great book. Dobell's description of the portraits of Leeuwenhoek is incomplete, for he does not do justice to all the works that survive. He never used his great talents in examining all the instruments left, nor in assessing the images they could generate. His suggestion that Leeuwenhoek may have been the very first to cut sections is not historically justified either, for Grew, Malpighi and − of course! − Robert Hooke were all well known to have done this before Leeuwenhoek, even if not as skilfully. And he seems to betray an embarrassing degree of bias in arguing against the theory that Leeuwenhoek might have had Jewish blood. He calls this 'a wild speculation' and adds that the evidence against it is overwhelming. Dobell adds:

> Mr Bouricius assures me that 'there were practically no Jews in Delft' at that date: and . . . nor did they hold any municipal engagements (from which they were debarred). Mr Bouricius, who speaks with authority, says: '*In geen geval was van Leeuwenhoek een jood*' − Leeuwenhoek was no Jew anyway.

Dobell goes further, by giving an unreliable rendering of the entry for Leeuwenhoek's birth in the Baptismal Register at the New Church, Delft, for 4 November 1632, as follows:

> 4. dito. 1. kint *Thonis*, vader Philips thonis zn, moeder Grietge Jacobs, getuijges Thonis philips zn, Huijch thonis zn, Magdalena, en Catharina Jacobs d^r.,

Dobell translates this in a way that entirely avoids any need to reproduce the surname 'Jacobs' – a common Jewish surname – by translating that entry in the following words:

> 4th ditto (i.e. November 1632). 1 child *Tony*, father Phillips, son of Anthony; mother Maggie, James's daughter. Witnesses: Tony Phillips's son, Hugh Tony's son, Madeleine and Catherine, daughters of James.

The Dutch translate the English name James into Jacobs, since the former does not exist in their language. But since in English we have both, there is little reason to transcribe the name Jacobs as anything but Jacobs: the name, of Jewish origins, is a familiar one to English speakers.

You do not need to be any distinguished scholar of the Dutch language to see the discrepancies between those two versions, though I think you would need to be a considerable authority before putting forward a definitively correct version. There is one modern literal translation that seems to me near-perfect. It was made for the English version of the official guide-book available to visitors to Delft entitled *Life and Work of Antoni van Leeuwenhoek* and was translated by a native Dutch-speaker, Mrs M. E. Adriaanse, from the original:

> 4. ditto. child Thonis, father Philips thonis' son, mother Grietge Jacobs, witnesses Thonis Philips' son, Huijch thonis son, and Magdalena, and Catharina Jacobs' daughter.

Whether the name *Huijch* should be changed to *Hugh* may be a matter of opinion, but the banishing of the very name 'Jacobs' by Dobell is less easy to forgive. There was a great degree of fashionable anti-Semitism in the 1930s in Britain as well as in Germany and elsewhere, and the fact is reflected in literature. However, this is an interesting example and it is revealing to find that such social codes – which are embarrassingly inept to a later generation – can apparently change the way the scientific record is compiled.

But the greatest 'deficiency' of the book is one which is inherent in its compilation, and one which Clifford Dobell himself specifically emphasised. It is that the theme – microbiology – was only a small portion of Leeuwenhoek's work. Dobell mentions several times that he is writing about Leeuwenhoek's work in the two fields in which Dobell was himself a specialist: bacteriology and protozoology (i.e. the study of bacteria and protozoa). Dobell wrote:

> Leeuwenhoek's observations on insects, rotifers, and a host of other 'animalcules' are equally remarkable; his researches on blood-corpuscles and the capillary circulation are already classics; his comparative studies of spermatozoa now stand as a landmark . . . while his other investigations in anatomy, histology, physiology, embryology, zoology, botany, chemistry, crystallography and physics, only await editors for their proper appreciation.

In spite of Dobell's appeal to understand his book as a selective treatment of just two aspects of Leeuwenhoek's work, it has rarely been seen in that light. So meticulously documented were many of the historical sections Dobell compiled that his book has been taken as the definitive summary of Leeuwenhoek's work and ever since the book was published in 1932 it has given Leeuwenhoek the reputation of being the father (or perhaps the 'midwife') of microbiology whilst diverting attention away from his work in those other areas of interest. I guess there might be five, perhaps ten, or possibly more volumes as expansive as Dobell's *Antony van Leeuwenhoek and his 'Little Animals'* yet to be written before we gain any real insight into the breadth of Leeuwenhoek's investigations. Take his role as a founder of plant anatomy. An account in the *Proceedings of the Royal Microscopical Society* for 1979 said: 'Leeuwenhoek used his superior lenses to look at plants; but are we to regard him as a major figure in botanical microscopy? Scarcely, one would think; his numerous missives to the Royal Society mark him as a dilettante . . .' As though in answer to that onslaught, a paper appeared two years later which analyses a vast number of important discoveries, supported by detailed drawings of the finest degree, which Leeuwenhoek made in the limited field of wood anatomy *alone*! Its author describes how Leeuwenhoek calculated the forces necessary to lift a column of water to the top of a 30 metre tree. When a

famous British botanist suggested that a feature Leeuwenhoek
observed in wood might be caused by the way he cut sections,
he set up an experiment to disprove it. Leeuwenhoek had
written: 'It may seem somewhat strange to you Gentlemen
Amateurs that I speak with such firm conviction . . .' and
added that his collection of different wood specimens had
already – by 1676 – amounted to 50 types. Most of the
conventional histories of the era explain that it was Marcello
Malpighi and Nehemia Grew who were the true 'fathers of
wood anatomy', but I have looked closely at Leeuwenhoek's
red pencil drawings in the Royal Society correspondence and
they are of a higher quality. He describes the cells found in
trees, the characteristic pits that mark the vessel walls in wood;
he argues against the belief that winter wood is more durable
and strong than summer wood, and demonstrates that the
belief only came about because the bark detaches itself more
easily from actively growing (i.e. summer) wood, resulting in
insect attack. He examines the claim that French ships and
casks are superior to those from Holland, and relates that to the
effect of higher temperatures and light levels on the growth of
oak in southerly latitudes. But he demonstrates that this is only
true of the hardwood trees, like oak; when softwood species are
grown in higher temperatures, Leeuwenhoek finds, they produce an inferior timber. All these findings he made with his
little microscopes; all of them are exactly confirmed by modern
timber technology. When he examined the structure of the cell
walls of *Myristica*, the nutmeg, he even seems to have observed
that the solid cell wall is built up of diagonal layers of minute
fibrils. This is the view that we have gained since the 1930s
through the use of sophisticated microscopy, and more recently by using the electron microscope. In that instance
Leeuwenhoek's observations were anticipating the work of
scientists almost two and a half centuries later!

He went on to show that nutmeg could act as a pesticide:

I was much surprised that I could not discover any mites
among the nutmegs. Therefore I placed about a quarter of
a nut among a group of mites, when I perceived they fled
from it. Moreover, I took a glass tube somewhat larger
than a swan's quill, one end of which I stopped with cork,
and after putting into the glass tube some hundreds of

mites, I cut a small piece of nutmeg of a size that would fit into it; and I perceived that the mites next to the nutmeg soon died. I then put another piece at the other extremity of the tube where there were many live mites, which also died in a short time.

To satisfy myself still further I took a glass tube, thirteen inches long, a half an inch in diameter. One end of this I closed by melting it, and put as many mites into it as I calculated would in volume equal half a cubic inch, and according to the nearest estimate I could make, they amounted to 150,000. After a quarter of an hour they spread themselves from the mass they were when first put in the tube, and spread all over the tube; I then split a very sound and good nutmeg into four parts, one of which I put into the open end of the tube so that I might see through the microscope what effect it would have on the mites when they approached it.

Most of them I saw creeping towards the open end of the tube, and when they came to within a straw's breadth of that part where the piece of nut touched to glass by two of its corners, many of them turned back, though they could have passed by the nut without getting any nearer than an eighth of an inch from the main substance of it. The retreat, as I might call it, of these multitudes of mites afforded a very interesting sight, for here it appeared that the evaporation or vapour from the piece of nutmeg was so noxious and offensive to them, that they drew back faster than they had advanced towards it, in order to make their escape from the tube.

Some others of the mites which had advanced so far as to be about a hair's-breadth distance beyond the nutmeg were soon arrested in their course and, losing all movement, they expired. Moreover, I observed numbers of mites creeping along the glass, near the part of the nutmeg covered by husk, and they would have escaped if I had not intercepted them by placing another piece in their way, so that they could not get out without passing the broken part of the nut. Hence it appeared to me that the vapour of the nut evaporates much more feebly through the husk than from the newly broken internal portion. Hereby, not only was the escape of the mites prevented, but all that were near the nut died there, and in the space of forty-eight hours, out of that great number of mites, only very few were left alive.

He carried out some practical experiments with the use of pesticides, for he showed that a little sulphur dioxide would soon kill grain moths in a glass jar, and when he set out to repeat the experiment by burning fresh sulphur he noticed that 'while I was preparing to burn the sulphur I saw them all lie dead, having been killed by the bare odour of sulphur which had been left in the glass.' He calculated that half a pound weight of sulphur would be sufficient to fumigate a local granary 24 feet long, 16 feet wide and 8 feet high which was infested with the moth larvae. After an experiment, two days later he returned to the granary and found a few moths flying around still. He concluded that they had probably been hatched from pupae that were already formed when the first treatment had been applied, and he added 'I am assured that so long as the moths are enclosed in their aurelia case or covering, the smoke of the sulphur can do them no injury.' He therefore recommended that the treatment be repeated when the dormant moths had hatched, so ensuring the sterility of the entire grain harvest.

Leeuwenhoek proved satisfactorily that an infestation was passed on by adult moths, rather than being naturally 'formed' out of decaying matter (the view of those adhering to the theory of spontaneous generation). In this, his poetic entrancement with the beauty of microscopical nature came to the fore:

Can any man in his sober senses imagine that the moth of which I have given this description, which is duly provided by nature with the means to propagate its species, furnished with eyes exquisitely formed, with horns, with tufts of feathers on its head, with wings covered with such multitudes of feathers all of different shapes, and they covering the wings in every part; can this moth, I say, adorned with so many beautiful features, be produced from decay? For in a word, in this little creature – contemptible as it seems to humans – there shines forth so much perfection and skill in formation as to exceed what we imagine in larger animals.

He examined mites, apart from in laboratory conditions as you might nowadays term it, by carrying them around in glass tubes and even had his wife do the same, so that bodily warmth would provide them with 'incubator' conditions. He kept cheese in closed tubes, and maggots in separate containers, to show that – no matter how offensive cheese might become – it did not produce maggots spontaneously, but only by the intervention of adult egg-laying insects. He opened out the 'nodous part of the liver of a sheep' and discovered flukes; he studied the parasites of frogs by keeping them in captivity and examining their excrement for the small opalinoid organisms they revealed. By blowing up organs with air, he demonstrated the pathways taken by vessels in animals he dissected for the purpose, and devised a series of experiments to prove that hair was pushed up from beneath as it grew, rather than (as was believed at the time) growing like a plant from the apex.

Leeuwenhoek studied conception by allowing rabbits and dogs to mate and then sacrificing them at intervals afterwards, studying where the spermatozoa had travelled, and what happened next. According to A. W. Meyer, writing in 1937, he is almost certain to have seen the early embryo – the blastocyst – which has been so much in the news following test-tube fertilisation experiments in human beings.

He corresponded with Robert Hooke, after the initial difficulties of their relationship, who proposed that Leeuwenhoek should be elected a Fellow of the Royal Society; and with a host of luminaries of the age. His influence was immensely strong throughout the eighteenth century and beyond. He died on 26 August 1723. The last few days of his life showed that he was still active in his mind, for he dictated letters to the Royal Society on 'corpuscles in blood and in the dregs of wine' and R. Boitet, the Delft publisher, wrote that Leeuwenhoek was dictating a report on the gold content of sand specimens (sent to him by the East India Company) only 36 hours before his death.

His 67-year-old daughter Maria wrote (or rather, dictated, for her shaky signature suggests she was not accustomed to writing) a sad and grief-laden little note to the Royal Society to accompany Leeuwenhoek's bequest of the little case of microscopes. She wrote:

Delft, 4th October 1723

Most excellent Sirs,

Instantly upon the sad death of my beloved father Anthonij van Lewenhoek I took care to have this my loss made known to you by our reverenced and most learned pastor, Peterus Griebius; adding to it that after a space of six weeks would be sent to the noble and far famed Royal Society, in London, a little cabinet with magnifying-glasses, made of silver wrought out of the mineral by my dear departed father his very self; which same is now sent to your Excellencies, just as my late father made it up, with twenty-six magnifying-glasses in their little cases: truly in itself a poor present to so celebrated a Royal Society, but meant to betoken my father's deep respect for such a learned Society, of which my most beloved and dear Father, of blessed memory, did have the honour to have been a fellow. Your most humble servant now begs Your Excellencies please to be so good as to let me have word whether this trifling gift has come safe into the hands of the far famous College, so that I may rest content that I have fulfilled my Father's wish.

Wherewith, most famous Gentlemen, your most re-spectful Servant and my father's Grief-stricken Daugh-ter now and hereafter will ever be, and remains,

Your humble Servant,
Maria van Leeuwenhoek,
Antoni's daughter.

Some months later in 1724 the Society sent a silver bowl, set with the arms of the Royal Society. Now this is missing, like the microscopes. By the end of the nineteenth century, Antony van Leeuwenhoek had been largely forgotten. A celebratory meeting to honour him was held in Delft during September 1875 to celebrate the bicentenary of his discovery of microbes. The date was wrong; and the site of Leeuwenhoek's 'house' was also incorrectly identified. The occupant, Mr J. B. A. Muré, took it all in good part, received the delegates with warmth, and agreed to have a plaque erected in Leeuwenhoek's honour attached to the premises. He undertook to be responsible for it in perpetuity.

The Royal Society, however, took no part in the celebrations. Not only did they omit to send a delegate to the meeting, but Dobell discovered that they did not even acknowledge that an invitation had been received from Holland. Within a few years all this had been forgotten even by the Dutch, and the house where the commemorative plaque had been installed was demolished.

Leeuwenhoek would not have minded that in the least. He worked for himself, and the advancement of knowledge in a field where – as he knew – there was nobody better from whom you could ask advice. He refused to teach others, or start any kind of microscope 'school', for he felt it would be unrewarding and he doubted the motives of those who might follow him. Individuality and creative freedom were what mattered above all to Leeuwenhoek. He declined to reveal his working methods, and always insisted that he had microscopes that were better by far than those he showed to visitors. In this respect, remember that the silver microscope collection he sent to London was made up of (by then) old microscopes with average magnifications. He knew perfectly well how easy it is to attract people whose interest in science is self-advancement through the channels of administrative power and personal rivalry, and once wrote:

> Most students go to make money out of science, or to get a reputation in the learned world. But . . . in discovering things hidden from our sight, these count for naught. And I am satisfied too that not one man in a thousand is capable of such study. Above all, most men are not curious to know: some make no bones about saying, 'What does it matter if we know this or not?'

I believe that Leeuwenhoek's views were in some ways superior to those that are popular even today in our ultra-learned world. Microbes play an important role in the ecological systems of our planet, and are co-inhabitants of our world with a purpose and a niche that is no less worthy than ours. Ever since the time of Pasteur, we have known of microbes as enemies of mankind and our 'unceasing war' against them has been seen as a fundamental task of our species. This, as I have said before, will not do. In the first place, many kinds of microbes promote health, and these forms – which I named *salugens*, or 'health-promoters' – must be more fully seen in contrast to the *pathogens*, or disease germs. I do not doubt that dosing ourselves or the environment with chemical agents against pathogens may be far less fruitful in the long term than harnessing the energies of these bounteous, beneficial microbial allies we so little understand. Secondly, the disease-causing types are a tiny minority of all the microbes that surround us. They are important, and our ways of exploiting plants and animals through agriculture, and our artificial methods of surviving in a civilised society, clearly make them significant organisms in the affairs of human kind. But we do not imagine that everyone we meet is a criminal murderer, simply because a tiny minority of people are, and it is irrational to view microbes as our enemies when the overwhelming bulk of them are working, directly or indirectly, for our good.

Leeuwenhoek has often been criticised for his failure to link bacteria with disease. Clifford Dobell wrote that, though Leeuwenhoek deserves every credit for not speculating ahead of the facts,

> nowhere in his writings does Leeuwenhoek associate entozoic protozoa or bacteria with the causation of disease . . . and it was left to others to elaborate his great discovery into the vast present-day *corpus* of medical protozoology and bacteriology. In a sense, therefore, he missed the great practical implications of his revolution.

In the same vein, Paul de Kruif said: 'Antony Leeuwenhoek failed to see the germs that caused human disease, . . . had too little imagination to predict the role of the assassin for his wretched creatures . . .' But I wonder if this apparent 'ignorance' was nothing of the sort. Ideas of contagion were rife before Leeuwenhoek's time, and the fact that he looked at his faeces when he had diarrhoea, for instance, makes one wonder whether at the back of his mind he did associate some of the animalcules with the disease all the while. If that is the case, why did he not press the matter? I do not believe for a moment it was because he was not interested in the possibility, but because he had a deeper and greater instinctive understanding of microbe life than our modern era can boast. You have been brought up to visualise microbes and germs as synonymous;

bacteria are to be avoided, microbial life in general is harmful.

This is an erroneous notion, and it has perverted attitudes to microbiology for a century and more. Microbes are our confederates, our neighbours, our partners, our servants. They are tiny but vital members of the world's community, and deserve intelligent investigation and understanding. The statistical minority which can cause disease cannot allow us to ignore that vital fact. If we can learn to co-operate with these living forces and utilise them for a carefully husbanded future life-style, then we will enter an unprecedented age of ecological harmony and true understanding.

Leeuwenhoek did not dwell on the possible malevolence of a few species of micro-organism, but concentrated instead on their abundance in nature and the entrancing activity of their own ways of life. This indicates that he was not only the first to observe microbial diversity, but was the pioneer of microbial ecology. Antony van Leeuwenhoek understood the wonder of microbes in a way modern science cannot equal – more than three centuries ago he embodied a breadth of vision and insight that we still need to inculcate in the scientist-of-tomorrow.

6 | Baker and the Tiny Flea

The simplest of all the early simple microscopes were the flea-glasses that were popular throughout Leeuwenhoek's life. The earliest type is pictured in a book by A. Kircher entitled *Ars Magnis Lucis et Umbrae* published in 1646, though later versions were similar. They consisted of a lens mounted in a pillar-based holder with a pointer on which the flea could be mounted. The essential idea was that the offending insect would be trapped by the fingers and its back would be impaled on the spike, so that the creature could be watched in its death-throes. I am not aware that there is any documentary evidence for this, mind, but if you try to catch a flea and impale it, this is what inevitably happens, and I cannot imagine that (in the more blood-thirsty era of the Regency time and the years before) the observer had found any humane way of dispatching the culprit before impaling it (Fig. 13).

The lenses magnified a few times, perhaps ten diameters, and gave a reasonable view of the insect. This kind of microscope is well known and widely documented, and its relatively poor optical performance may be one of the reasons why the simple microscope began life with a 'bad press'. The Dutch microscope manufacturer, Samuel Musschenbroek of Leiden,

Fig. 13 A flea-glass designed to be stored under a thimble-shaped cover when not in use. The soda-glass lens of focal length 1 cm showed the insect clearly, though such lenses were of insufficient quality to make serious study possible. They were popular around 1700.

joined in the business later by his young brother Johann, refined this design by making a jointed arm for the support of the specimen, a clear improvement on the bent wire which was the supporting structure in the basic flea-glass. The ball-and-socket joints in the Musschenbroek dissecting microscope are known as '*Musschenbroek nuts*' to this day. Samuel Musschenbroek made simple microscopes for the great pioneer of insect anatomy, Jan Swammerdam, who flourished in this field between 1663 and 1675. The high-power microscopes he made were of a quite different nature, with a wooden body and a relatively complex means of holding the specimen adjacent to the lens. There was also a sliding aperture plate which could give a range of circular perforations, with which the user could carefully regulate the cone of light that fell on the specimen under examination. The most elaborate microscope of this era must be the example produced by J. Langlois of Paris before 1720. Not only did the microscope have a focusing control that was nearly two inches (50 mm) in diameter, but it was heavily embossed with floral ornamentation and must have taken a considerable time to manufacture. A somewhat simplified version of this was used by the French biologist Louis Joblot, Professor of Mathematics at the Royal Academy of Painting and Sculpture, Paris. Joblot's version was made by Le Febvre, Manufacturer of Mathematical Instruments (according to the

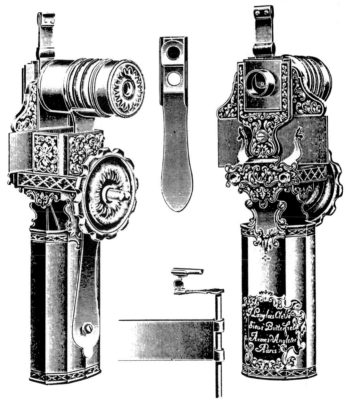

FIG. 14   An elaborately decorated simple microscope produced for Louis Joblot, Professor of Mathematics at the Royal Academy of Painting and Sculpture, Paris, around 1720. He stated that his instruments were made by Le Febvre; the engraving on this model reads:
'J. Langlois, Eleve du Sieur Butterfield, au Armes dAngleterre Paris'

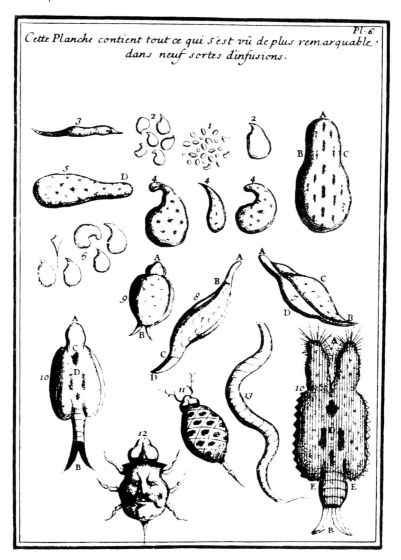

Pl. 6

Cette Planche contient tout ce qui s'est vû de plus remarquable dans neuf sortes d'infusions.

FIG. 15 Many of Joblot's drawings, like these pond organisms from his *Descriptions et usages . . .* (1718), show forms that are hard to reconcile to their actual appearance. Leeuwenhoek's studies were of a far higher observational quality.

description published in his *Descriptions et usages de plusiers nouveaux microscopes tant simples que composez* of 1718).

Louis Joblot carried out some important work with his instruments, including the demonstration that micro-organisms did not appear in an infusion that was sterilised by boiling, and kept in a closed container; but he was a poor observer. His drawings of rotifers, which he named 'aquatic pomegranates', show them as almost unrecognisable, grotesque shapes that have little in common with the real thing. Leeuwenhoek's drawings and descriptions were of a far higher standard, and would do well as modern illustrations. Though it would be unhelpful to take a sideways step in the argument, and consider Joblot as a microscopist in great detail, it is worth noting that his demonstration that sterilised culture liquids would remain free of microbes is the kind of revelation that most people might associate with the Pasteur era one and a half centuries later – yet it was discovered by using the simple microscope.

Perhaps the most widespread example of this kind of instrument was the compass microscope which was popular in the middle years of the eighteenth century. The design was simple enough: a handle supported a central strut on the top of which was fixed the single lens. An arm carrying the specimen was hinged near the base of the strut, so that the distance between the lens and the specimen could be altered at will until the image was in focus. The term 'compass microscope' was appropriate since the little instrument was quite similar to a geometrical compass of the kind used in school for drawing circles; with a lens where the 'pencil' might be and a specimen holder instead of the 'point'. The best type of compass microscope had a fine screw thread attached to a focusing knob between the two arms. By turning this the object: lens distance could be more accurately varied. In some ways the compass microscope was a technological version of a flea-glass, and because of the relatively inefficient focusing mechanism compass microscopes were not much used for serious investigation. As amateur toys they were popular, and as instructional aids they must have played their part; they were not part of the mainstream of interest in microscopes and microscopy.

Yet the compass microscope did popularise one interesting device which has its counterpart in modern microscopes – the so-called Lieberkühn reflector. The way a Lieberkühn worked was exceedingly simple, yet effective – the reflector was mounted with the lens in the centre, facing *away* from the observer. When the object was held up to the light, rays were reflected from the Lieberkühn on to the solid specimen, so that

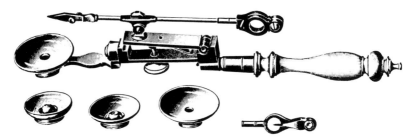

F<small>IG.</small> 16   Some versions of the compass microscope were finely constructed of brass, steel and ivory. The silvered Lieberkühn mirrors surrounding the four lenses in this example were used to reflect light onto an opaque specimen.

the side facing the observer was brightly lit. The effect is to produce a beautifully illuminated image, in which the details of the specimen are lit evenly – in the manner of the most expensive ring-flash illuminators obtainable today. Quite why the name Lieberkühn should be attached to the device is unclear: Leeuwenhoek himself published a design for what looks like a reflector of this sort, though there is no evidence he ever made one; and the design is clearly visible in René Descartes seminal book *Dioptrique* which was published in 1637. The idea existed a century before it became popular through the compass microscope. Modern instruments often have a separate light-source mounted high in the body, which shines down through the objective lens on to the specimen. This is how metals are usually illuminated by the microscopist. But Lieberkühns are also found in daily use, and some accessory manufacturers specialising in photographic equipment still sell them. The Lieberkühn is a useful device, and I am surprised there are not more of them around.

The kind of microscope that became the mainstay of eighteenth-century microscopy – at least for the study of inanimate objects – took the form of a short barrel through which the specimen could be slid, mounted in a protective holder; the lens was then brought to focus by screwing it in or out of the barrel as the observer held the device up to the light. This form of microscope was named after James Wilson, who began to make them at the beginning of the eighteenth century, and is known to this day as the Wilson Screw-barrel microscope. You have seen how the Leeuwenhoek microscope was actually a design quoted earlier by Hooke, and that the

Lieberkühn was actually promulgated long before the man of that name was even dreamed of; so it will come as no surprise whatever to know that the Wilson Screw-barrel microscope was certainly *not* invented by Wilson. The earliest examples of screw-barrel focusing were found in compound microscopes, and an Italian inventor named Tortona first described the principle to a meeting of the Physics Mathematical Roman Academy of 1685, according to the Oxford historian and anatomist Savile Bradbury. The application of the same principle to a simple microscope was the idea of Nicholaas Hartsoeker, one of Leeuwenhoek's great rivals – Dobell refuses to quote anything Hartsoeker wrote as a mark of profound dislike – and that dates from 1694. Some observers have since accused Wilson of plagiarism, but his own writings show that he was not claiming to have invented a new kind of microscope at all, when his first screw-barrel instruments were announced. All he said was that he proposed to market a microscope 'made by James Wilson' which was of a recently invented type. There was no overt emphasis of who the originator was, true, but then there was no deliberate attempt to mislead anyone into thinking that Wilson was claiming to be the originator, as well as the producer (Fig. 17).

So much for the instruments – who was to be their new champion? We might heed the conclusion of Lorande Loss Woodruffe, a Professor of Yale University in the earlier years of this century, who in 1918 wrote: 'Among the English disciples of Leeuwenhoek, it was Henry Baker, of London, on whom seems to have fallen the Dutch microscopist's mantle . . .' though, for good measure, Woodruffe had the sense of perspective to add, 'though, it is true, considerably reduced.' Baker was born on 8 May 1698 and, after schooling, he was apprenticed to a bookseller. He was appointed tutor to a deaf-mute child when he was 22, and soon found ways to teach lip-reading to the child. So considerable was his success that he took on other deaf-mutes from reasonably wealthy families, and established a school for them. He taught them how to follow conversation, how to draw and write, and supplemented all this by taking them out on 'life-experience' walks during which he would teach them about the everyday features of London life. This blend of the theoretical and practical brought him considerable success, and he kept his system a closely guarded secret lest the security it offered be diluted. He made a considerable fortune from his activities, even levying a bond of £100 which the families of all his pupils had to pay as a guarantee of secrecy. In his spare time Baker wrote florid verse, including a poem that had far-reaching influences in some respects!

You may know of a fragment of verse that runs as follows:

Great fleas have little fleas upon their backs to bite 'em,
And little fleas have lesser fleas, and so *ad infinitum.*

These lines are often seen as a light-hearted encapsulation of the dawn of microbiology. They were taken from a book entitled *A Budget of Paradoxes* published in 1872 by Augustus de Morgan, a mathematician with a humorous bent (his favourite target for attack was the natural countryside, for he claimed to hate meadows and above all, trees; whilst adoring smoky steamboats . . .). The sense of that poetic couplet is well known even though frequently misquoted. Many people now assume that it was nothing more than an up-dated variation on the theme used by John Donne in his own poem, 'The Flea':

Marke but this flea, and marke in this,
How little that which thou deny'st me is;
It suck'd me first, and now sucks thee,
And in this flea, our two bloods mingled bee;
Though know'st that this cannot be said
A sinne, nor shame, nor losse of maidenhead.
Yet this enjoyes before it wooe,
And pauper'd swells with one blood made of two,
And this, alas, is more than wee would doe.

This, you could argue, uses the pre-microbiological concept of a flea as an influence, a law unto itself, a breaker of human rules, but none the less liable to the same physical constraints of nature as anything else; whereas the lines on the flea written by Augustus de Morgan in 1872 reflect the insight of what elsewhere I have called 'microscopic consciousness', and the smallness of the flea becomes instead a mirror of more weighty philosophical implications, no matter how lightly expressed. But that is not the case, for the Morgan lines are a plagiarism of an earlier poem. Those quoted above were lifted and vulgarised from a sharper work by far, under the title *On Poetry*, and published in 1733 by Jonathan Swift. The extract that concerns us here runs as follows:

> So, naturalists observe, a flea
> Hath smaller fleas that on him prey;
> And these have smaller fleas to bite 'em,
> And so proceed *ad infinitum*.
> Thus every poet, in his kind,
> Is bit by him that comes behind.

This doggerel verse is a clever, almost delicious, parody of a work entitled 'The Universe' and published in 1727 with the subtitle 'A Poem Intended to Restrain the Pride of Man'. The author was – Henry Baker. It ran, in part, thus:

> Each Seed includes a Plant: that Plant, again,
> Has other Seeds, with other Plants contain:
> Those other Plants have all their Seeds; and Those,
> More Plants, again, successively inclose.
> Thus, ev'ry single Berry that we find,
> Has, really, in itself whole Forests of its Kind.

And it concluded with the revelation of 'Reason's piercing Eye', that Adam's loins must have contained 'his large Posterity, All people that have been, and all that e'er shall be.' Baker added: 'Amazing thought!' as a thoughtful postcript to the purple prose of his insightful analysis.

　　If imitation is the sincerest form of flattery, parody must imply notoriety. Henry Baker had risen from being a teacher of deaf-mutes into a leading writer on contemporaneous microscopical methods, the author of a best-selling book on the microscope, and a fellow of the Royal Society in 1741. It is said that his work as a teacher (surely that, rather than his florid 'poetic' writings) brought him to the attention of Daniel Defoe in 1727, who invited him to his home. Two years later, Baker married Defoe's daughter, Sophia. He carried on a lengthy correspondence with scientists in many places, travelled widely, and was personally responsible for introducing into Britain the Alpine Strawberry and the medicinal rhubarb.

　　His account of microscopy provides a fascinating insight

into current beliefs. It is given in two books, the second being in many ways an 'appendix' to the first. The full title of his initial opus, quoted from the second edition (the preferred copy of this work, for it contains additions not ready for the first) is:

THE MICROSCOPE Made Easy:
OR, I. The *Nature*, *Uses*, and *Magnifying Powers* of the best kinds of MICROSCOPES
Described, Calculated and *Explained*: FOR THE
Instruction of such, particularly, as desire to search into the WONDERS of the *Minute Creation*, tho' they are not acquainted with *Optics*. Together with
Full Directions how to *prepare*, *apply*, *examine* and *preserve* all sorts of OBJECTS, and proper cautions to be observed in viewing them.
II. An Account of what surprizing *Discoveries* have been already made by the MICROSCOPE: with useful Reflections on them.
AND ALSO A great variety of new *Experiments* and *Observations* pointing out many uncommon Subjects for the Examination of the CURIOUS, by HENRY BAKER, Fellow of the *Royal Society*, and Member of the Society of *Antiquaries*, in *London*. Illustrated with Copper Plates. The SECOND EDITION: With an additional *Plate* of the *Solar Microscope*, and some farther accounts of the POLYPE.

It received its imprimatur in 1742 and was published the following year by Robert Dodsley of Pall Mall. Later editions came out in 1743 (the revised second) and then, without changes, in 1744, 1754, 1769, 1785 . . . as well as translations and overseas editions. It is interesting to glance through the pages, to see the impression that is given of the state of microscopy at the time. Baker refers to Leeuwenhoek's microscopes as being the simplest possible, and then goes on to describe five main kinds of instrument that were popular. First is the Wilson Screw-barrel: 'The first that I shall mention, is Mr. WILSON's *single Pocket Microscope*; the Body whereof is made either of Brass, Ivory, or Silver.' Said Baker, the magnifying glass you intend to use is screwed in until the object 'fits the eye', which you would soon know, because the image became 'perfectly distinct and clear.' The second instrument was a version of the first: the scroll-mounted screw-barrel microscope in which the instrument was held on a curved metal bracket, the light being reflected up through the optical axis by means of a mirror. The larger instrument was clearly more stable and easier to use, and according to Baker it 'may be made to answer almost all the Ends of the large *Double Reflecting Microscope*, which I shall presently describe.'

The double microscope was a compound instrument, a modification by Scarlet and Marshal of the tripod-mounted Culpeper microscope, explained Baker, 'than which it is less cumbersome, may be managed with much more Ease, and by means of a reflected Light, is capable of shewing Objects in a clearer and more pleasing Manner.' He then described a *Solar, or Camera Obscura Microscope*, which 'is composed of a Tube, a Looking-glass, a convex Lens, and WILSON's single Pocket Microscope before described.' Just as Baker says, the instrument is a screw-barrel microscope through which the sun's rays could be directed. Not only did this give a small angular source, which produced a singularly contrasty and 'hard' image, but the brightness was such that the image could be projected on to a wall, and the struggling flea would thus provide a marvellous spectacle for a large group of spectators. A louse, he added sagely, could be made to appear as long as five or six feet, and the screen for the image could be made up in a roller that could be pulled down 'like a large Map', like the slide-projector screen of today.

The difficulty with the solar microscope was that the small angular size of the illuminant tended to produce diffraction fringes around the image; but these were probably more than offset by the dramatic impact of the microscopic image. Solar microscopes must have made many converts to the cause of microscopy. Finally he described a *Microscope for Opake Objects* – the compass microscope. The use of the speculum mirror, said Baker, would provide a wonderfully illuminated image 'and who ever trys it,' he added, 'will I believe join in my Opinion, that he never before saw an opake Object with so much Clearness, and in so perfect and true a manner.'

Henry Baker's writings are full of interest, though they have been misunderstood. Clifford Dobell describes him as a copyist and compiler, who drew largely upon Leeuwenhoek's work for his own purposes. But this is unfair. Baker freely drew on many workers, including Hooke, Swammerdam and Henry Power the physician and naturalist, and he frequently and enthusiastically acknowledged his indebtedness to these great men. Without Baker, it is difficult to see how Leeuwenhoek's name would have remained alive, for the Dutch pioneer was largely forgotten in England – and in Holland – within a short while following his death. Baker shows he was an eager disciple of everything Leeuwenhoek wrote, and that is no lamentable attribute. It is equally interesting to see that Woodruffe, in his examination of Baker's book, quotes from it at length and explains that his books 'exploited the compound microscope'. Now, as I have said, the compound microscope has so come to dominate our thinking that it is easy to make statements like that, almost without realising it. But Baker's book did nothing of the sort. Of the five types of instrument he described, which you can put at six if you include his examination of Leeuwenhoek's own instruments, only one was compound (the modification of the Culpeper type); all the others were simple, single-lensed instruments and it was these, *not* the compound microscope, that were the mainstay of Baker's writings. Interestingly enough the descriptions of the simple microscopes are replete with little references to the ease of use, the beauty of the image, the clarity of the focus; and the only type that comes in for implications of inconvenience in use is that tripod-mounted compound microscope.

By this time, the first kind of 'sliders' were in use. Today we are familiar with microscope slides as slips of glass based on the Imperial standard of three inches in length and one in breadth, but these were not introduced until the mid-nineteenth century, when good quality glass began to become available. The mounting system that was popular during the eighteenth century (for it was purpose-built for microscopes of the screw-barrel type, with their sprung stage) was the slider, made of ivory or bone. Each was perforated with a series of about four countersunk holes, and slips of mica or talc cut to fit and held in place with circlips enabled the user to hold his specimen firmly in place. The presence of several specimens in one slider was merely a means of economising on space and effort. Sliders became the norm for all microscopists mounting dry specimens, and once again it is interesting to note that it was the sprung-loaded stage most widely found in *simple*

Fig. 17 The screw-barrel microscope, shown here with five ancillary lenses and eight sliders (one is in position in the barrel), was popular throughout the eighteenth century. The spring-loaded specimen stage was not suitable for the examination of larger pond organisms, such as *Hydra*.

microscopes that led the way. One famous engraving of a slider in use appears in Bonanni's book of 1691 entitled *Observationes*, and this has been unwisely described as 'the first slider' in one recent history of section-cutting. It is certainly not the 'first' – a well-drawn example appears, for instance, in the 1686 Italian work *Nuove inventione di tubi ottici*, which pre-dates the previous example by five years, and it was probably demonstrated in Rome during the previous year.

One can glean an impression of the kinds of specimen that were in vogue from consulting a descriptive catalogue of the time. One of the earliest examples, accompanying a Wilson Screw-barrel microscope sold by Culpeper, gave them as:

No 1: Hairs; Down of a Moth's Wing; Dust of the Sun Flower; Dust of Mallows.
No. 2: Scales of a Sole Fish; a bit of Sponge; the Pith of Elder; a bit of Cork; a bit of the Pith of a Rush; the leg of a Moth; and a bit of Feather; the Eye of a Fly; the Wing of a Fly.
No 3: A Louse and dead Flies; the Down of a Thistle; a bit of Human Skin; a bit of Wasps Nest; a bit of Leaf Gold; the Hair of the Head; and a small Spider.
No 4: A spare Slice for Liquids.

These were supplied with a brass slider with slips of glass 'to put any small living Object into at pleasure.' It is interesting to note here the presence of elder-pith and cork, reminders of Hooke and Leeuwenhoek.

Later, sliders were also made of hardwood and some fine collections of reference specimens – different types of timber, for instance – still exist. The slider in due course gave rise to the slide, since it seemed easier by the early 1800s to obtain rectangles of glass rather than to cut circles of mica. Specimens were then mounted between two pieces of glass which were covered with gummed paper. For much of the Victorian era, glass slides were covered with decorated gummed paper, with an aperture cut out for the specimen to be examined, a convention that seems to stem from the appearance of sliders. By the end of the nineteenth century, it was becoming popular to mount specimens in a clear resin such as turpentine or balsam from pine-trees, which not only aids their preservation but provides optical benefits too.

It was the limitations imposed by this form of mounting specimens that projected the development of the microscope forward with a leap. The reason is that, though a slider is obviously suitable for any small dry specimen, and could do as well for any minute object mounted in water, it is not applicable to *larger* living organisms. Insects that are to be examined in the round pose no great problem, for they were usually held either on the point of a specimen pin – just as Leeuwenhoek used to do – or else they were gripped in forceps made for the purpose. Most specimen forceps were made with the blades of the forceps at one end of a metal rod which tapered to a sharp point at the other. The point was used for impaling insect specimens, and was normally guarded when not in use by an ivory disc. The disc was usually left white on one side, but black on the other, so that a suitably contrasting background could be chosen for small objects that were temporarily lain against the surface. What with forceps, ideally adapted for holding objects that needed a low magnification, and sliders, perfect for any small thin object, the choice of mounting method would have seemed flexible. For most specimens it was. But there was just one type of organism which suddenly came in for attention. When Henry Baker began to study this creature he found that none of the available methods was suitable for bringing the organism under the view of his simple microscope lenses. That creature was the *polyp* . . . and the need to investigate it efficiently and easily eventually gave us the shape of microscopes that has come down to us today.

The polyp, as Baker and his contemporaries knew it, is a small freshwater organism known today as *Hydra*. An adult *Hydra* is a centimetre or so long and thicker than a human hair, so it is clearly visible to the naked eye. However, a microscope is necessary before the details can be seen. *Hydra* was known long before Baker's work; it was documented in some detail by Leeuwenhoek who studied it in glass tubes filled with pond water. He left good descriptions of it with beautifully observed diagrams. He even described a tiny louse parasite that lives on the *Hydra* itself!

At the time of its discovery by Leeuwenhoek the life-style of *Hydra* attracted no interest whatever. By the time the eighteenth-century biologists returned to the subject, their revelations of the same details Leeuwenhoek had observed fell on a more alert and attentive audience, and the interest generated was intense. *Hydra* is a simple organism, a tiny tube of two layers of cells with a closed lower end, which is fixed as a rule to the substrate, and an upper extremity that divides into fine tapering tentacles. These are armed with extraordinarily sensitive stinging cells, *cnidaria*, which can shoot a barbed thread through passing prey and ensnare a small water-flea or something similar for food. The prey is drawn towards the mouth of the *Hydra* and is then taken into the body cavity which balloons out to accommodate it.

Baker included many details of his work on *Hydra*, explaining that this 'insect' – it is in fact a coelenterate – was first discovered by Mr A. Trembley 'who now resides in Holland'. Baker explains:

> The Polype has eleven Horns or Arms, and adheres by the Tail to a little Twig. Their common Posture is, to fasten their Tails to something, and then extend the Body and Arms into the Water: and they make use of their progressive Motion to place themselves conveniently for this Purpose. Their Arms are so many Snares, stretched out to catch small Creatures in the Water: and when any Insect happens to touch an Arm it is caught, and conveyed to the Mouth by the contracting of that Arm, or if the Creature struggles the other Arms assist.

Baker concludes that the descriptions 'seem to mean the same *Animalcule* Mr. Leeuwenhoek describes,' and he adds that Trembley saw the reproduction of *Hydra*: 'The *Polype* brings forth its Young from the exterior Parts of the Body, and not always a single one at once . . . the more one searches into the reproduction of these Insects, the more evidently it will appear to be done by true Vegetation.'

This is quite true, for – though *Hydra* does produce testes and ova from time to time, and does indulge in a sexual phase of reproduction – its normal method of procreation is vegetative. A small protuberance forms on the side of the body, and over a day or two this slowly elongates until it produces tentacles of its own. The daughter-*Hydra*, as it is known, separates itself from the parent body and moves off to live an independent existence. *Hydra* can creep slowly sideways, but for movement over greater distances in the water the organism reaches out with its tentacles, releases hold with the base of the body, swings over to the new site of attachment, and eventually releases the grip with its tentacles once more – a form of cartwheeling in slow motion.

Leeuwenhoek's description of the reproduction of *Hydra* was written with his characteristic clarity:

> I discovered a little animal whose body was at times long, at times drawn up short, and in the middle of whose body a still lesser animalcule of the same type seemed to be fixed by its hind end. When I saw such a little animal attached to a bigger one, I imagined that it was only a young animalcule attached by chance to a big one; but by finer attention to the matter, I saw that it was a reproduction . . . after a lapse of sixteen hours I saw that its body and its horns had increased in bigness, and four hours later still, I saw that it had foresaken its mother. When I discovered the young animalcule aforesaid, I also perceived that, on the other side of the body of the first animal, there was situated a small round knob which I did see getting bigger, from time to time, for the next few hours. And at last it appeared to be a little pointed structure, which had so far grown in bigness in the course of thirteen or fourteen hours, that you could make out two little horns upon it. After the lapse of another twenty-four hours, this last

mentioned animalcule had four horns, whereof one was small, a second a bit bigger, and the other two much bigger; and these last the little animal stuck out at full length, or pulled in short. And another three hours later this little animal was gone off from his mother.

At one time or another I let the draughtsman have a look at the horns as they were being stretched out, or later pulled in, and with me he was obliged to exclaim 'What wonders are these!'

At the time, the idea that animals might reproduce asexually or vegetatively was unheard of, indeed undreamed of. Leeuwenhoek's graphic description – as vivid as any you would find – should have triggered a flood of excited interest. That would doubtless have been the case if science moved in predictable, logical pathways. In fact scientists are as prone to fashion and to trendiness as anyone else, and since nobody was ready for such revelations they were simply ignored. By the time Trembley and Baker were writing on the same subject almost half a century later (Leeuwenhoek's description was dated 1702) the minds of men were better attuned to such unlikely observations, and on that prepared ground the concept flourished at last.

Being a relatively lowly animal, *Hydra* not only has this capacity for vegetative reproduction, but a wide-ranging ability to repair itself when damaged. As Baker wrote in 1743:

Cut a *Polype* where or into what Parts you please, transversely, each Part becomes a *Polype*. [Mr Trembley] first cut one into four Quarters, and let them grow; then divided each Quarter, and proceeded, subdividing, till he obtained fifty out of one: and he still has by him several Pieces of the same *Polype* thus cut above a year ago, which have produced Numbers of young ones. If a *Polype* be cut the long Way, through the Head, Stomach and Body, each Part is half a Pipe, with half a Head, half a Mouth, and some of the Arms at one of its Ends. The Edges of these half Pipes gradually round themselves, and unite, beginning at the Tail End; and the half Mouth and half Stomach of each becomes compleat.

Henry Baker expanded these observations into a book of 222 pages, costing Four Shillings Bound, under the title *An Attempt Towards a Natural History of the Polype* and published in 1743. He wrote that 'That curious Observer of Nature, Mr LEEUWENHOEK, first took notice of this Animal, and the uncommon way its young are produced, in the Year 1703 . . .' and went on to list a series of experiments which show how many were the ways he could envisage for mutilating an adult *Hydra* – yet from which it would recover. The subject of regeneration has considerable ramifications through biology, and yet for one and a half centuries his interesting demonstrations were forgotten. The experiments were these:

I.	Cutting off a Polype's Head;
II.	Cutting a Polype in two Pieces, transversely;
III.	A Polype cut into three Pieces Transversely;
IV.	Cutting the Head of a Polype in four Pieces;
V.	Cutting a Polype into two Parts, lengthways;
VI.	Cutting a young Polype in two Pieces whilst still hanging to its parent;
VII.	Cutting a Polype lengthwise through the Body, without dividing the Head;
VIII.	A Repetition of the foregoing Experiment, with different Success;
IX.	Cutting a Polype in two Places through the Head and Body, without dividing the Tail;
X.	Cutting off half a Polype's Tail;
XI.	Cutting a Polype transversely, not quite through;
XII.	Cutting a Polype obliquely, not quite through;
XIII.	Slitting a Polype open, cutting off the End of its Tail;
XIV.	Cutting a Polype with four young Ones hanging to it;
XV.	Quartering a Polype;
XVI.	Cutting a Polype in three Pieces the long way;
XVII.	An Attempt to turn a Polype, and the Event;
XVIII.	Turning a Polype inside out;
XIX.	An Attempt to make the divided parts of different Polypes unite;
XX.	A speedy Reproduction of a new Head;

XXI.   A young Polype becoming its Parent's Head;
XXII.   A cut Polype producing a young One, but not
        repairing itself.

There seemed no end to Baker's invention! *Hydra* itself is an organism ideally suited to this sort of manipulation, and related experiments with flatworms such as *Planaria* have been useful in more recent years in helping us to understand the way repair of tissues takes place and this in turn assists the unravelling of healing in higher animals.

This is where we encounter the practical problem that was to lead to the next phase of development in simple microscopes. It is this: how do you examine *Hydra* with the types of simple microscope then available? Leeuwenhoek's design for an *aalkijker* was no longer available, and it certainly will not fit on to a compass microscope. It is too large to fit into the cells of a conventional slider, for an adult *Hydra* is too long to fit into the space available. If Baker had used compound microscopes as his instrument of choice, then there would have been no problem. Indeed he must have used such an instrument for many of his studies, because of the limitations that we have just considered. But he must have been dissatisfied with what they could do, for he determined to find a new type of microscope that *would* accommodate his 'polypes'.

The books Henry Baker wrote were usually sold by two dealers, M. Cooper and J. Cuff – and Cuff was already well known as a manufacturer of optical instruments. No doubt Baker would have discussed his difficulties with Cuff, and in 1750, Cuff began production of a small portable microscope which overcame the problems. Instead of screwing the lens into a barrel of metal or ivory (as was the case with the screw-barrel microscope) or into one arm of a compass holder, the lens-holder fitted into a horizontal arm mounted above a circular stage. Light was directed through the device by means of a mirror mounted near the base of the instrument. The first of the instruments stood free on a disc of brass, but this was found to be unstable and by the time the first 35 or so had been made, Cuff was designing them to be clamped to a block of wood as a firm base. Because these microscopes could be used for watch-glasses containing pond organisms, they had a use that extended into areas that the screw-barrel type of microscope could not reach. They were so well-matched for observing aquatic organisms that they soon became known as *aquatic microscopes*.

Some idea of the difficulties that the earlier microscopes caused can be gained by looking at the *Hydra* specimens featured by Baker in his book. The organisms featured on Plate IX of his *Microscope Made Easy* reveal several interesting features. His Fig. III, for instance, is clearly a redrawn version of Leeuwenhoek's own illustration – and, in a drawing as in literature, it loses something in the 'translation'. The Leeuwenhoek example comes from Fig. 4 of the illustration to his letter dated 25 December 1702 (published in *Philosophical Transactions* No. 283, 1703) and Baker says as much in his account (p. 95, para. 2) where he attributes his description to 'the substance of Mr. Leeuwenhoek's Letter to the *Royal Society*'. In fact Leeuwenhoek described eight tentacles on the *Hydra* he examined, though his own artist drew in nine; but Baker restores the originally intended number of eight in his own version. His diagram published as Fig. VIII is a good drawing of *Hydra* at its healthiest – long and fully-expanded tentacles, three daughter-*Hydras* developing vegetatively, all the signs of an active adult individual. His higher-power studies in Figs IV (showing a young bud) and V (showing a well-formed vegetative daughter-*Hydra*) are good drawings of normal organisms. But Fig. VI, portraying an adult in the partly-contracted state, and Fig. VII, in which a similar *Hydra* is viewed from above, are very different. If you look at them casually, they seem to show the appearance of a *Hydra* that is in the halfway stage between expansion and contraction.

But the tentacles lie tapering and bunched together; that and the conical outline of the body, reveal one important thing: the organisms in these two figures are dead or, at best, dying. If you examine this organism on the stage of a modern microscope – or on the stage of an old-fashioned compound microscope from Baker's day, come to that – then the appearance of *Hydra* is easy to study under its normal, life-like conditions. A *Hydra* damaged enough to give the appearance shown in Baker's Figs VI and VII has been abused. It is almost as though the organism was crying out for a less cramped environment, a technique of observation that did not produce such dire effects on the specimens themselves. I do not want to

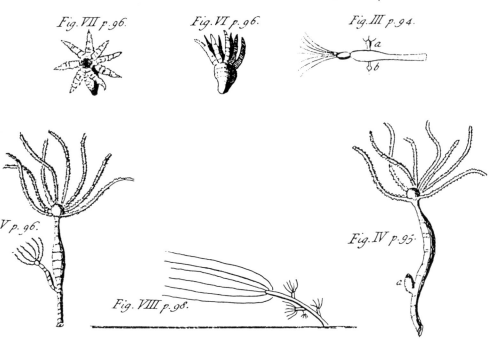

FIG. 18   Baker's drawings of *Hydra* show forms (Figs VI and VII) that appear harmed by early mounting methods.

go so far as to suggest that it was 'Hydra Lib' which led to the development of the aquatic microscope. But I have no doubt that, without the upsurge of interest in these tiny animals in the eighteenth century, the instrument would not then have been invented. We are taught that the technical perfection of an instrument leads to a desire to use it, and to the opening up of new areas of investigation. Here we have the converse – a new 'area' itself causing the development of the instrument – if you like, the organisms imposing their own design demands upon the instrument-maker. In that sense, we owe a lot to *Hydra*.

7 | The Microscope in the Field

Students of microscopical history know aquatic microscopes by a fuller title – 'the Ellis Aquatic Microscope'. As you might by now expect, the instrument was not invented by John Ellis at all. Apparently he encouraged Cuff to modify slightly his earlier design of an aquatic instrument, so that the bracket holding the lens could be turned round horizontally through 360° and also slid in and out, nearer to the main body of the microscope, or further away from it; so that the whole area of the stage could easily be examined. Some writers have insisted that this was the design's cardinal fault, since the slightest movement of the microscope would make the polyps contract. This is unlikely, firstly because the first aquatic microscopes (designed for the study of these organisms) did not have this feature, and secondly because the amount of movement imparted to the instrument by moving the lens arm sideways would be quite enough for them to detect – and not that much less than the amount of movement caused by gently moving the glass trough containing them from side to side. Furthermore, *Hydra* is not as sensitive as all that. Preparations of the living organisms can be gently moved around without causing them all to contract into little spheres. Since *Hydra* can accept small movements of its surroundings without responding overmuch, and moreover since small movements of the lens arm are transmitted to the organism in either event, I question that widely quoted conclusion. In my view, the aquatic microscope was a simple design by Cuff, based on Baker's recommendations drawn from practical difficulties encountered during the use of conventional single-lens microscopes. It seems likely that the swivelling lens bracket resulted from manufacturing convenience.

Cuff emphasised that this new design of microscope made it possible to examine all kinds of specimens – sections, minerals, entire insects, blood and seminal smears, *and* larger living pond organisms like *Hydra* – by issuing in 1758 a pamphlet entitled 'The Description of a Double and Single Microscope Very Convenient to View All Sorts of Objects', which showed an aquatic microscope on to which a compound body could also be fitted, if desired. Six years later, John Dollond, the prominent microscope-maker, issued a pamphlet describing 'The Aquatic Microscope as Improved by John Ellis F.R.S.', and this showed a microscope with a main supporting pillar bearing a circular stage and a pivoted mirror beneath it, with a lens arm at the top of the instrument into which the lens-holders could be screwed. The needs for portability and stability were ingeniously met by having the microscope pack down into a box, on to the lid of which the pillar was mounted during periods of use.

The great Swedish naturalist Carl Linnaeus (1707–1778) used a simple microscope in his work of classifying the living organisms of the known world. The family name Linnaeus was taken by his father when he entered the University of Uppsala but he adopted the name of Carl von Linné on being ennobled in 1761. Linnaeus laid down the principles of modern classification, using the binomial system of two names, the genus and the species, which are familiar as the so-called 'Latin names' of biology. Not only did he systematise classification, but he gave us the unambiguousness on which taxonomists (i.e. classifiers) depend. A robin or a blackbird cover several different species, depending upon where you live, whilst a plover, a lapwing and a peewit are all different names for the same species. Of the names Linnaeus proposed in the eighteenth century, roughly 4,400 species of animals and over 7,700

FIG. 19   An aquatic microscope marked by Cuff of London was used by Carl Linnaeus of Uppsala. It unscrews when not in use to stow into a fish-skin covered wooden case. Focusing is carried out by sliding the lens holder up or down in its friction mount.

species of plants still remain current; in each case the name bears the suffix 'L.', citing Linnaeus as the authority for the name. Our own species for example is *Homo sapiens* (L.). His *Species Plantarum* of 1753 together with his *Genera Plantarum* (5th edn, 1754) has been accepted by botanists as the starting-point for botanical nomenclature in general, just as his *Systema Naturae* Vol. I (10th edn, 1758) is the international reference for zoologists.

Linnaeus regarded the simple microscope as a normal and everyday part of his essential equipment. His ingenious approach to the classification of plants, for example, centred on the study of the sexual reproductive organs of each species – for flowering plants, this meant the structure of the stamen and pistil, the exact order of the flowering structures, rather than such irrelevant criteria as leaf shape, height or habitat. Many of these structures are hard to discern with the naked eye, and a portable microscope would have been necessary for much of his work.

In fact Linnaeus's microscope is preserved in the house attached to the old botanic garden at Uppsala, which he occupied during his professorship, and is now a Linnaeus museum. It is kept in a closed cupboard alongside his own writing desk and near the fur-covered bed in which he died. The microscope fits into a neat case covered with polished fish-skin in the manner of many eighteenth-century micro-scopes. It is made of brass and has a mirror, a circular stage, and a lens bracket typical of the aquatic microscope genre. The main supporting pillar is engraved with the words 'CUFF, London'. Two lens-holders are present, but one of them now lacks the glass lens. The remaining lens has a low magnifica-tion of only $\times 20$, but as the diameter of a lens is decreased so its magnification increases, hence the smaller diameter of the missing lens shows that it would have been considerably more powerful.

The fact that Linnaeus used a microscope at all was overlooked for many years, indeed the *Journal of the Royal Microscopical Society* for 1905 (p. 253) published a letter from Sir Frank Crisp, a noted microscope collector, saying that he 'never heard that Linnaeus did' though he had since read a short sentence by Linnaeus in Latin indicating his use of a Cuff microscope for the examination of fungus spores. No doubt he used the high-power lens – now missing – for that work. However, this microscope cannot have been made until around 1750.

As noted above, in 1732 Linnaeus set out on his celebrated journey from Uppsala to Lapland, collecting, observing and classifying everything he saw. His journal, which is preserved in manuscript at the Linnean Society in London, records that he set out on 12 May 1732, old style, at eleven o'clock, 'being at that time within half a day of twenty-five years of age'. His reference to carrying a microscope occurs in his introductory remarks and, since his account paints such a vivid picture of his preparations for the journey, it is interesting to have it in full:

> Having been appointed by the Royal Society of Sciences to travel through Lapland, to investigate the three kingdoms of Nature in that country, I prepared my clothing and other necessities as follows. My clothes consisted of a light coat of West Gothland cloth without folds, having small cuffs and a collar of shag; leather breeches; a round wig; a green leather cap, and a pair of top boots. I carried a small leather bag, half an ell [i.e. 30 cm] long, but somewhat less in breadth, furnished on one side with hooks, so that it could be shut fast and hung up; this bag contained one shirt; two pairs of false sleeves; two nightcaps; an ink-horn, pencase, microscope and spying-glass; a gauze cap to protect me occasionally from the gnats; this journal, a parcel of paper stitched together for drying plants, both in folio; a comb, my [manuscript] Ornithology, Flora Uplandica and Characteres generici. I wore a short sword at my side, and carried a small fowling-piece between thigh and saddle, an eight-angled stick graduated for the purpose of measuring. My pocket-book contained a passport from the Governor of Upsala, and the Society's recommendation.

The drawings Linnaeus brought back did not feature any high-magnification studies. His manuscript journal, for instance, shows a kind of cranefly that he labels alongside '6, Culex ibidem . . .' – *gnat from the same place* – shown hardly magnified at all, but with the veining of the wings excellently portrayed. The same figure is published in the translation of his *Tour in Lapland* edited by James Edward Smith, then president of the Linnean Society of London, in 1811. Smith amends the caption in his edited version to give Linnaeus's chosen name of this species, and omits Linnaeus's 'No 6' from the manuscript: 'The annexed figure represents a large kind of gnat caught in the same place (*Tipula rivosa*).' This species is now known as *Pedicia rivosa* (L). I imagine that Linnaeus used a lens for this drawing, even though it is not much larger than life. However, so far as microbes were concerned, Linnaeus had a blind spot. The few names he proposed for micro-organisms were sometimes sound, but his general understanding of their systematic relationships was poor and even behind the times. On the other hand, it is clear that – without his simple microscope – Linnaeus's work would have suffered. He used it to good effect in his studies of higher plants, and in some cases of animal species.

There is one curious instrument associated with Linnaeus. It has a Latin inscription saying it was given to Bernard de Jussieu in memory of the pleasant association he had enjoyed with Linnaeus in Paris during 1738. The instrument is a compound microscope with a body of paste-board, and it came to light in 1887 in the lumber room of the Harmony Society, a German community at Economy, Pennsylvania. Ten objective lenses were found with the microscope, which has no trace of a maker's name and is of strange appearance. Why it was presented and how Linnaeus came by it are unknown. It may be that he gave it away, as an instrument he did not need – since his later Cuff microscope was a single-lensed instrument – or he may have purchased it especially for Jussieu

FIG. 20   A page from Linnaeus's journal, which he kept during his journey of 1732, bears this sketch of *Tipula rivosa* (now known as *Pedicia rivosa*), probably undertaken with the low-power lens which still survives with his microscope (Fig. 19). The venation of the insect wings shows with remarkable clarity.

as a token, in response to some indication from the recipient that it would be a suitable gift. It would be tempting to conclude that it was the instrument he took with him to Lapland in 1732 (just six years earlier) but it is not in any way a portable microscope, and there are no indications that he planned sufficient microscopical work to make it worth while transporting anything so bulky. Perhaps it was a microscope which he used in Paris, in the continuance of his studies. But it must be said that compound microscopes of this sort were more decorative than functional, more so where serious and detailed scientific work was in view.

Some simple microscopes were not capable of use as high magnification microscopes, but were made purely for dissection. Of these, one of the best known is the botanical microscope invented by William Withering (1741–1799) late in the eighteenth century. It took the form of a basic little microscope mounted into the lid of a box, with a chain to pull it into the upright position as the box itself was opened. The instructions read:

> When the [floral] parts in question are very minute, and require a nice and careful dissection, place the microscope upon a table, and raise it, if necessary, on a book or two, so that the eye may be applied with ease immediately over and close to the glass (b). Lay the object to be examined on the dark stage (c) until you see the object upon the stage perfectly distinct. With the needle in the left, and the knife in the right hand, the elbows resting on the table, proceed in the dissection at the same time that the eye is applied to the glass (b).
>
> When the microscope is shut up, the instruments and the hand-glass are to be put into the cells destined to receive them, and the whole forms a shape and size convenient to carry in the pocket.

The lenses were not of fine quality, and had a limited magnification. It is partly because of microscopes like Withering's, which doubtless had a use for low-power work in the field, that single-lensed microscopes came to be viewed as dissecting instruments. They were nothing of the sort – as we will see, they reached surprising (and, today, scarcely recognised)

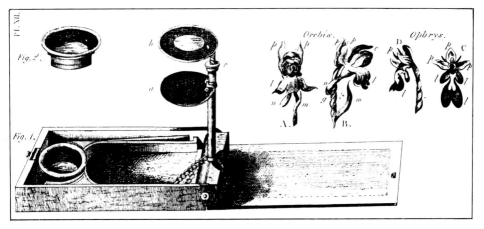

FIG. 21   William Withering's pocket microscope was a true dissecting instrument, and was not intended for high-power magnifications. It is not well designed; a hand-lens might have been better for field use.

optical performance. I am inclined to think that Withering was not entirely as convinced of the value of his microscope as his advertisements for it suggest, either: many of the illustrations in his book on the British flora – published under the title *Withering's Botanical Arrangement* – were surely carried out with a better instrument by far.

William Withering himself came from a herbalist's family: his father was an apothecary and young William was sent to Edinburgh School of Medicine. The then Professor of Medical Chemistry, William Cullen (1710–1790), was a prominent physician, a specialist in herbal medicine who had introduced many such remedies into practice with considerable success. Withering graduated in 1766 and seemed unlikely to make his mark in botany: in spite of his background, his upbringing and his training, he seemed ill-disposed towards botany. When he was told by John Hope, Professor of Medical Botany, that he might well win a gold medal if he did well in botany, he still wrote that even such an inducement did not make the subject seem attractive. He disliked botany intensely. Yet, within a decade of graduating, he had published the first edition of his *Botanical Arrangement* in 1776.

The reason for this flowering of interest was a woman. Withering fell in love with Helena Cookes, an artist whom he met through her father after moving to set up practice in

Stafford. Apparently she loved to draw flowers of every description, and William Withering realised that an assiduous study of her favourite subjects would be useful in attracting her attention. In the meanwhile all he had been taught about herbal remedies seems to have temporarily left his mind – by the time the *Botanical Arrangement* appeared he was safely married to Helena, and the book itself represented a major advance in systematic botany without any reference to his earlier devotion to the medicinal uses of the plants. Withering apologised for the omission: 'Many people will be surprised to find so little said upon the medicinal virtues of plants but those who are best enabled to judge of this matter will perhaps think that the greater part of that little might well have been omitted . . .' Besides which, the delectable Miss Cookes, artist, had been interested in the plants, not their properties . . .

After his marriage to Helena he took up a position in Birmingham, and was for thirteen years physician to Birmingham General Hospital in addition to having a practice of his own. He became friendly with James Watt and Matthew Boulton, both members of the Lunar Society, was an active member of the Society for Promoting the Abolition of the Slave Trade, and carried out research to disprove the existence of the mythical phlogiston – the heat-producing entity that was theorised to be produced by combustible agents, in the years before the activities of oxygen were proved. In his later years, Withering had the honour of being sent plant specimens by Thunberg – Linnaeus's successor at Uppsala – and receiving his agreement to some changes that Withering wished to incorporate in the third edition of his *Arrangement*. He died of consumption in 1799, survived by his two children William junior and Charlotte. For all his caution and scepticism Withering was responsible for the introduction of digitalis in the treatment of heart disease, through the careful records of effective dosages in his book *An Account of the Foxglove and some of its Medical Uses* (1785). It was a vital step which has saved countless lives since – indeed the leaves of the foxglove plant are still used as a standard medicine in many areas. As the great botanist lay dying, witty commentators were observing: 'the flower of physicians is indeed Withering.'

He, undeniably, was; but the simple microscope was flourishing and was now set for its final, greatest, days.

8 | The Image of the Single Lens

It is easy now, with hindsight, to look back and reflect that the early microscopists were merely marking time until the modern achromatic compound microscope was invented. Microscopists in the seventeenth and eighteenth centuries observed objects with the best instruments they had, aware that they had limits, and living within them as far as they could. We take hi-fi for granted now, without realising how the scratchy and decidedly *low*-fi discs of the 1930s were greeted with enthusiasm for their remarkable realism. It is so easy to take digital watches for granted that we lose sight of the difficulties our forebears encountered when designing an accurate mechanical analogue watch that would merely keep reasonable time. And how did we ever manage with clanking engineering adding machines, slide rules and books of logarithm tables in the days before the ubiquitous calculator arrived? Only retrospective convenience allows us to sneer at those pioneering efforts, and to marvel at the assumed sophistication of our own era.

This must be kept in mind when we go back to the first years of the nineteenth century. Then the difficulties of using tiny single lenses were well known, and the limits they imposed were accepted as the norm. So, as the simple microscope reached its zenith, it was not as a primitive and useless device used as a last resort, nor was it regarded as a low-power instrument with severe limitations. The simple microscope was an everyday tool of science. It was carefully made, conscientiously used. Lenses work by refraction, and as we have seen they depend on the fact that a beam of radiation – light, sound, or anything else – changes direction if it passes at an angle from one medium where it travels at a given velocity, into another medium where the velocity is changed.

With that in mind, the way a lens is built can immediately be perceived. A lens is shaped so that the light rays near the centre are hardly refracted at all, whilst those at the periphery are refracted very much more. The result is that the original parallel beam of light we began with is focused to a spot. If there is a positive lens in the room where you are reading this book – spectacles for someone who is long-sighted, a magnifying glass, even a lens supplied with a micro-type book – you can use it to project an image of the window or the lights against any suitable surface. The image is inverted, of course, but none the less it is a striking demonstration of the focusing action of a positive – magnifying – lens which most people never try. The distance between the centre of the lens and the image of a distant object is known as the focal length of the lens. The stronger a lens, the shorter the focal length; and the shorter the focal length the greater the magnification. Most people with normal sight hold an object about ten inches away from the eyes, so if you were to use a lens with a focal length of *five* inches, that would enable you to have the object only half the distance away – so it is effectively said to be magnified *times two* – usually written ×2. The microscopists of the last century used to use this convention in talking about lenses, so that a lens with a focal length of 1/10 inch had a magnification of 10 inch ÷ $\frac{1}{10}$ inch = ×100, and one with a focal length of 1/40 inch magnified ×400. Nowadays we use a standard image distance of 250 mm instead, which makes us duly metricated, but adds to the difficulty of calculation (thus the latter example above becomes 250 mm ÷ 0.625 mm = ×400).

There are many other physical constraints but there is no need to deal with them here. Some, like the Rayleigh limit, matter more to lens designers; others, like numerical aperture, relate to the theoretical limits of a lens to resolve fine detail and are not always the same thing in practice. Resolution is a concept one must embrace, at least in principle; it is defined as the smallest separation between two self-luminous points that a given lens can still show as two distinct entities, and not one. In practice it simply implies *how small are the details you can see with the lens*. This matters more than magnification. For instance, the resolution of detail in a printed picture in a newspaper is a function of the half-tone screen used in making the block. If there is an object in the printed picture that cannot quite be 'resolved' by the screen, then there is no hope of seeing it more clearly by using a magnifying lens on the printed page. If the detail is not resolved, it never will be by that system.

The reason why I am not going to dwell on this topic is because lenses often do not fit the theory. In practical terms, the finish of the lens and its actual shape introduce incalculable modifications into the equation. Thus the lens that Leeuwenhoek fitted to his microscope now at Utrecht was blown in a flame and does not have the normal spherical shape of a modern ground lens. This means that its resolution is higher than you might expect, and the region of the image which seems sharp and clear to the eye is wider than it would be if the lens had been ground normally. I have taken photographs using early nineteenth-century lenses of similar theoretical performance, only to find them very different in practice – one gave a sharp and crisp image, whilst the other had some optical deficiencies which gave a perfectly useless result. In my work on these old lenses, though one has to rest part of the evidence on the physical parameters of each lens, I have tried to keep my findings rooted in the experimental realm, rather than the theoretical. What matters most is how an image looked, not what it *should* have been, and I have relied on this pragmatic principle throughout. Remember that – when it comes to what can actually be seen by a lens – Leeuwenhoek's hand-made one proved superior to a Zeiss doublet lens, carefully constructed with the wisdom of the nineteenth century.

There is an additional limitation on lens performance that all theoreticians seem to imagine exerts considerable adverse effects on single lenses which simply does not! This is the chromatic aberration. We have seen how a beam of light can be brought to focus. But the degree of refraction to which light is subjected depends on its wavelength. The greater the wavelength, the less the light is refracted. This means that in theory the red light rays in a beam of white light will be brought to focus further from the lens than the shorter wavelengths of blue light. This limitation was a perpetual annoyance to users of early compound microscopes and was the main reason why the simple microscope was preferred. The main practical benefit of the compound microscope was far more mundane – its size! The length of the body meant that the instrument could stand on a table and present itself at a convenient height for the eye. At the same time, a notebook alongside was at a comfortable distance from the observer, so that he or she could look down the microscope, and then across at the page to record the observations, without discomfort. Simple microscopes either had to be held in the hand, or else stood on a table at such a height that the observer was forever bending down to look through and then having to half straighten up again before making a note or completing a drawing. In addition, scientists love to show off their possessions. This certainly explains why the compound microscopes of the late nineteenth century were sometimes festooned with luxurious devices that nobody ever needed, and why the earlier versions were so elaborately covered with designs and superfluous adornments. One microscope made for King George III by the London manufacturer George Adams in 1760 is sculpturally complex and so difficult to use that I doubt whether it was ever regarded as a scientific instrument at all – it was surely a clear example of showing-off!

The problem with compound lenses is that, unless they are constructed with a full understanding of refraction, and contain balanced arrays of lenses which cancel out, then the aberrations are magnified as much as the object. This was so until around the middle of the nineteenth century, when compound lenses became generally available and were of reliably high quality. Since that time the simple microscope has been superseded, and its reputation besmirched in the process. It is important to realise that the best modern microscope is only capable of results that are perhaps four times better than a Leeuwenhoek lens. If you compare a modern car, a plane, a telescope or a camera with their earliest counterparts then you will find that the modern version is tens or even hundreds of times faster, or better, or more powerful, than the pioneering version. But a Leeuwenhoek microscope gives an image that is not much worse than that provided by a modern instrument. Here is a modern description of an image generated by a current microscope: 'The magnification (×50) might have been a typical one in the late 1700's, although present-day definition is likely to be superior', it reads. And here is another description from *Three Centuries of Microbiology*:

> Eighteenth-century microscopes were primitive instruments, limited in their usefulness by the chromatic and spherical aberration of their lenses ... The resulting blurred, multi-coloured image enticed many a scientist to believe that the microscope was useful only in creating artefacts.

Such statements are erroneous. A high-powered single lens produces an image that is acceptable for any normal purpose. It is not 'blurred'; it is not imbued with rainbow-hued haloes; it is not the distorted and fuzzy, indistinct image that everyone seems to expect. A single lens produces an image that compares favourably with the demands of much modern microscopy, and is often better than the poorly set up modern microscope which may be too complex for an untrained user to adjust properly. Chromatic aberration does not generate garish red-and-blue fringed images in these lenses; at its worst it may seem to give some minute components a lilac or orange tint they do not have in reality, but the effect is hardly noticeable. Spherical aberration is a defect of focusing due to the fact that the surfaces of lenses are usually spherical in shape. Many scientists imagine that this means that the centre of the image is sharp, whilst the rest is not; but in fact the effects of spherical aberration affect the whole of the image, and result in a change in magnification between the edge of the image and its centre. This too is of marginal importance in single lenses of relatively high magnification. We have become so used to the grandiose compound microscope, and are so used to the efforts that have been put into overcoming the problems of earlier compound systems of lenses, that we assume the simple microscope to have been indescribably bad. It was not. The limitations of the lenses in these early microscopes are slight, and in my view this kind of instrument would be ideal for use in student teaching, in field trips, and for third-world laboratories today. It is our laziness in wanting a microscope that is easy to

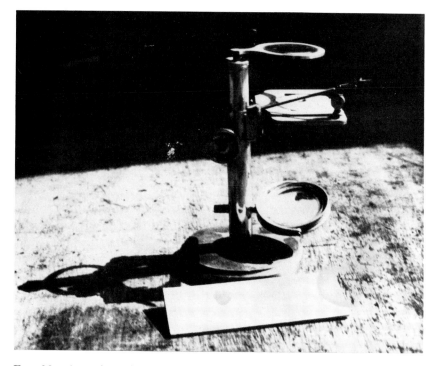

Fɪɢ. 22 A neglected botanical microscope in a museum collection. Not only is it mis-attributed as a 'low-power instrument' but it is wrongly assembled. The circular fitment mounted where the lens should be, at the upper extremity of the microscope pillar, is in fact part of a swing-out stage. Compare this with a refurbished specimen in Fig. 28.

use and comfortable, coupled with our instinctive modern tendency to admire anything that looks complicated and grand, which has allowed us to *assume* that the early microscopes were so poor. You may rest assured that, out of a million biologists who will tell you that an early nineteenth century microscope generated a poor image, hardly anyone will have ever been given a chance to look down one.

As you will know by now, I cannot imagine that anyone is even basically 'educated' without knowing what life is like down a microscope. On the other hand, I am equally aware how hard it is to provide a sense of scale and a frame of reference to a person not accustomed to looking down microscopes. Here are some concepts that may help to provide that sense of scale. Most microscopists think in terms of four levels of magnification. There are macroscopic views, followed by low-power microscopy, then high-power microscopy, followed by oil-immersion. What those mean in practice is as follows.

(a)   Macroscopy

This is the term for what photographers call 'close-up' work. We are talking of magnifications around ×10. At this level, all we do is make clearer structures that are largely familiar to us already. The largest cells can be seen this way, for instance. Some cells can be glimpsed by the naked eye – and I am not thinking of the giant cells of some seaweeds, a meter or so long, nor of such exceptions to the rule as an unfertilised hen's egg, which is also a single cell. No, the cells of which plants and animals are made can sometimes be seen with the naked eye. Leeuwenhoek's specimen, elder pith, is an example. I have already explained that if you break off a young stem of elder, *Sambucus nigra*, you can cut a fine wafer from the pith with a razor-blade and see the cells plainly, looking a little like polystyrene foam. If you look closely at the end-grain of polished wood you will see the tiny openings of the xylem vessels that conduct sap up from roots to leaves. These too were single cells. The hairs that are packed around the stamens of the popular garden plant *Tradescantia virginiana* (a plant we will hear more about later) and its hybrids are made of giant cells that can be seen with the unaided eye. The largest cell of the human body is the ovum, and that is visible to the naked eye as a minute whitish speck a little smaller than this full stop.

(b)   Low-power microscopy

Here we are talking about magnifications of ×50 or so. Larger pond organisms are well displayed by this level of magnification; so are the anatomical details of plant and animal sections. Microbe life begins to emerge, though you must not expect to see blood cells clearly and bacteria are still invisible. Macroscopy – if we can call it that – was the realm of the flea-glass, and low-power microscopy was the province of the early compound microscopes. You will recall that Robert Hooke worked at this order of magnification for his investigations for *Micrographia*.

(c) High-power microscopy

Here we are talking about magnifications of ×300 or so. Bacteria can be seen, all types of cell are easily visible and so are the details of structure within each cell – such as the nucleus and even the mitochondria that actually power the cell's life-processes. High-power microscopy is sufficient to enable us to see all but the minutest details. It is perfectly adequate to observe disease-causing bacteria, for instance; indeed high-power microscopy is sufficient for all the normal purposes of optical microscopical investigation. And this is the important point – *magnifications of this sort have existed for centuries through the single-lensed microscope!* Leeuwenhoek's best surviving lens provides ×266, and it has been claimed that his highest magnification might have been ×500. That may be over-optimistic, but it is equally unlikely that one of nine surviving instruments out of a total production of several hundred is going to be the most powerful of them all. If a lens of ×266 exists, then it is more likely that more powerful lenses have been lost. The most powerful single lens I have ever seen was produced by a British lens-grinder and approached ×1000 (though its resolution was no better than a lens magnifying half as much) and a lens magnifying nearly ×500 was identified in a collection of early nineteenth century microscopes in Holland some years ago, as we shall see. In fact lenses magnifying a little under ×200 were very common during that period, and this is enough for most biological specimens.

(d) Oil immersion

The obvious relationship between focal length and magnification means that, for the highest magnifications, the objective lens has to be as close as possible to the object. With an oil immersion lens system a drop of oil with similar optical properties to glass is introduced between the lens and the object, effectively producing an 'oil lens' that is virtually in contact with the object. The idea itself is more ancient than many people realise: indeed Robert Hooke mentioned a form of oil immersion as long ago as 1678.

> If you would have a Microscope with one single refraction, and consequently capable of the greatest clearness

and brightness, spread a little of the liquor upon a piece of looking-glass plate, then apply the said plate with the liquor next to the globule, and gently move it close to the globule, till the liquor touch; which done, you will find the liquor presently adhere to the globule, and still adhere to it though you move it back a little; by which means, this being a specifique refraction not much differing from glass, the second refraction is quite taken off, and little or none left by that of the convex side of the globule next the eye.

(Hooke, R., *Lectures and Collections*, II: pp 98–99, 1678)

Oil-immersion lenses provide the best results theoretically obtainable, and magnifications between ×600 and ×1000 are usual. At this level of magnification, resolution becomes near the 0.2μm limit imposed by the nature of light (where 1μm – one micrometer – is a thousandth of a millimeter). It is important to realise that a single lens magnifying around ×300 has a resolution of approximately 1μm, and you can actually discern fine detail that is smaller than this. For instance, I have obtained clear photographs with the Utrecht Leeuwenhoek lens of fine structures only $0.75\,\mu$m in thickness, even though its resolution is stated to be $1.3\,\mu$m: a lens can be used to *visualise* structures that are beyond its theoretical capacity for *resolution* – another reason for my use of the lenses to obtain working images, rather than relying purely on hypothetical considerations.

Into the above framework we have to fit the size of the objects that microscopists wished to study. Plant cells are often $50\,\mu$m in diameter, animal cells more usually 20μm. Smaller bodies include blood cells and the nuclei within cells, which measure 7–$8\,\mu$m; whilst bacteria vary considerably in size, but are usually in the range of 2–$5\,\mu$m. This means that almost all the above groups of features can be studied with lenses magnifying ×100 or more, and only a slight improvement on that is necessary to observe bacteria clearly. A survey by P. H. van Cittert in Utrecht during the 1930s showed that to have a resolution of $2.5\,\mu$m a simple lens of the type available between 1700 and 1830 would need a magnification of ×100; a projection (or solar) microscope – p. 101 – would have required a lens magnifying ×120 to obtain the same resolution, whilst a compound microscope of the period 1720–1820 would have needed a magnification of more than double that – ×250 – to give the same result. On this basis two conclusions can immediately be drawn:

(1) that the compound microscope was inferior to its simple counterpart;
(2) that single-lensed microscopes were capable of revealing most of the significant structures that can be resolved by a modern optical microscope.

In the following chapter I will show how these facts were given practical realisation by the microscopes of the early nineteenth century. For the present, one can only hope that the notion of degraded and indistinct images, of inadequate magnification and limited resolution, and of primitive and impractical construction, can be removed from our attitude towards simple microscopes. Not only were they admirable instruments produced by people of intelligence and dexterity, but could produce excellent results for their period. It has long been taught that the rise of the achromatic compound microscope in the mid-nineteenth century led to the discoveries in microbiology and the mushrooming awareness of microscopic structure that followed. This is a mainstay of historical dogma, a bedrock on which modern perspectives are founded.

It is wrong. The overwhelming 'breakthrough' discoveries were made before that era. When Leeuwenhoek's acquaintance van de Graaf studied the follicles in the ovary which bear his name, he used a simple microscope. So too did C. E. von Baer when he announced the discovery of the human ovum in 1828, at the same time that Robert Brown in England was unravelling the complexities of reproduction in plants. Though there was an explosion of interest in the microscope when it became more complex and more interesting, in the middle of the nineteenth century, many of the fundamental observations in microscopical biology had already been made. Indeed, most of the important discoveries – including the role of pathogenic bacteria in causing disease – which were later made, could have been carried out using the lens of Leeuwenhoek which survives to this day in Utrecht. The capacity of the simple microscope has been overlooked, and it is unfortunate it has remained so widely misunderstood for so long.

9 | Robert Brown and the Nucleus

Tiny particles suspended in a liquid will not keep still, no matter whether they are minute bodies from living cells, bacteria, or specks of carbon suspended in water; under a good lens they can be seen to be in a continual state of agitation. To the eye of the tyro they might almost look alive, but jostling organisms – watched for a little while – show each actually heading somewhere whereas these tiny inert particles oscillate about the same place, seeming to jump about incessantly. The effect is found in all particles less than a couple of micrometers in size, and the smaller the particle the greater the movement. When first seen it was assumed by some to be a sign of life – the tiny specks of moving matter were named by Robert Brown *active molecules* and it was imagined in some vague way that they were the hidden components of life. The phenomenon is now known as Brownian movement (sometimes as Brownian *motion*) and it is a popular demonstration to students of physics. Diluted Indian Ink shows the effect well, and so does milk – the suspended oil droplets show the unceasing flickering movement clearly. Gases manifest the same property, for small smoke particles can be seen to oscillate in a similar fashion. The effect of Brownian movement is produced by the ceaseless agitation of molecules, which bombard the particles on all sides with kinetic impulses. It was first noticed in the early days of microscopy – one ambiguous reference suggests that it may have been witnessed by Leeuwenhoek – though not until 1905 was the phenomenon fully explained in a paper by Albert Einstein.

The name of this property of fluids commemorates the name of Robert Brown (1773–1858). Here too, the name is anomalous, for it was not Brown who first observed it. As he pointed out in his writings, he did not claim that his observations were the first in the field and he cited two earlier workers who had witnessed the phenomenon. It deserves to be more widely known that it was Brown who named the nucleus – the essential 'brain' inside the living cell in which resides the genetic code which determines the characteristics of each living organism. Nuclei were first seen by Leeuwenhoek, for they are visible in some of his studies of blood cells from fish, but Brown first realised that they were found in a whole range of living cells.

Robert Brown was a Scot. His father was a Episcopalian minister named James Brown, and Robert inherited his intellectual rigour and sturdiness of personality. Young Brown was sent to the Marischal College in Aberdeen and later the University of Edinburgh to study medicine. He had been born in December 1773 and by 1795 enlisted in the Fifeshire regiment of Fencibles as Ensign with the duties of Surgeon's Mate. For the next five years he devoted all his spare time to academic study. He learned German, making a special study of its complex grammar, made copious notes about plant life, studied the works of such important earlier biologists and medical men as Sir Hans Sloane, and collected plants whenever he could. His break came in 1798 when he was on a visit to London. He had made the acquaintance of José Correa da Serra, at the time exiled from his native Portugal, and through his friendship was introduced to the influential President of the Royal Society, Sir Joseph Banks. Banks had voyaged with Captain Cook in his circumnavigation aboard the *Endeavour* in 1768–1771. Banks's house in Soho Square contained a rich library and a vast herbarium, partly derived from his own travels, and the two began an important friendship. In 1800 the Admiralty planned an expedition to chart the coast of New

FIG. 23   Robert Brown observed the cell nucleus; the ultramicroscopic effect of Brownian movement, cytoplasmic streaming and a host of other microscopic phenomena. This portrait hangs in the Linnean Society of London, and was painted by H. W. Pickersgill, R.A.

South Wales and New Holland, later named Australia, under the command of the young Matthew Flinders, then 27. At the time NSW and NH were thought to be separate islands. Asked to advise on the choice of a naturalist aboard Flinders's ship, the *Investigator*, Banks recommended Brown. With little delay, Brown obtained leave of absence from his regiment and made preliminary studies of Australian plants before joining the ship. He was Flinders's age, 27, and with an Admiralty salary of £420 was already a wealthy professional botanist. By the time Brown returned to England in 1805 (five years before Flinders, who was interned on Mauritius by the French governor and did not return until 1810) he had almost 4,000 species of plants together with a large number of geological and zoological specimens and copious notes and drawings. For the next few years he worked classifying the material – indeed, by the time Flinders returned he had already described 2,200 species of which 1,700 were previously unknown.

In that same year, 1810, Banks's librarian and assistant, Jonas Dryander, died, and Brown took his place. He worked as librarian and curator of the collections at Soho Square until Banks's death in 1820. Banks bequeathed to him the life tenancy of the Soho Square house together with an annual allowance of £200, with the stipulation that on Brown's death the priceless botanical collections and library were to pass to the British Museum. Robert Brown offered the collections earlier than this, in 1827, on condition that he became head of the botanical department at the Museum. During the following year the lengthy move took place, and it was in this manner that the first British national botanical collection was established. Brown had as his own assistant a surgeon named John Joseph Bennett, who in due course acted as Brown's executor after his death in 1858.

Brown had been elected a fellow of the Royal Society in 1810, and a fellow of the Linnean Society of London in 1822. He became the Linnean Society's President from 1849 to 1853. His travels took him across Europe and he became well known in Germany and France. Meanwhile Brown made important discoveries with the simple microscope that showed how important – and how effective – this forgotten instrument can be.

The detailed descriptions by Brown and the equally detailed and exquisite drawings by Ferdinand Bauer made on Flinders's voyage, coupled with the evidence of Brown's earlier work on the mosses, indicate that he had been familiar with the minute structures of primitive plants, and was almost certainly using a simple microscope prior to 1810. The Proteaceae was a group of plants laid down by Jussieu (one of the great old families of Lyons – it was to Bernard de Jussieu to whom Linnaeus presented the compound microscope mentioned on p. 117).

In devising a classification for the Proteaceae, Brown

wrote that microscopical examination of the shape of the pollen grains would be a sensible way of studying the relationships of the species. He added that pollen, 'I am inclined to think, not only from its consideration in this family, but in many others, may be consulted with advantage in fixing our notions of the limits of genera.' This is an interesting idea in several ways. It established the value of pollen grains in classification, and this in turn led to their use in the study of vegetational changes through the ages. The scanning electron microscope has given a new means of examining pollen grains closely, but it is worthy of note that it was the simple microscope which laid the foundations for the technique of pollen grain study.

In 1826 Brown noticed what proved to be a vitally important feature of the way conifers reproduce. To modern biology, this group of trees are included in the *gymnosperms*, a name meaning 'naked seeded' plants since the ovary lacks the covering which is normally found in flowering plants. This is the fundamental biological distinction between the coniferous plants and the flowering plants. Brown examined the reproductive structures of these plants and concluded that 'the ovarium was either altogether wanting, or . . . imperfectly formed' in the conifers. The importance of the observation was only realised later – but what I find so impressive is that Brown could so confidently and objectively reach his conclusion in the way he did. These are tenuous structures which modern microscopical methods enable us to demonstrate with relative ease. But to the true pioneer working on untrodden ground the task of interpretation becomes immeasurably greater. Brown showed the same kind of independent sure-footedness when in June 1827 he examined pollen grains of *Clarkia pulchella* (a popular garden annual) and saw inside each grain a large number of minute particles which seemed to be in continual motion. He named them *active molecules* and set about examining the pollen of many other plants to see if the same phenomenon occurred. The immediate impression might be that the 'molecules' represented the movement of some kind of ultimate living process. Brown seems to have wondered whether this was the case, for he moved on to examine old herbarium specimens, each more than a century old. In pollen from these old, dried specimens he saw the same phenomenon when the pollen grains were mounted in a droplet of water. Many of his predecessors would have been tempted to tell their friends: 'Look at this! Inside the pollen grains – the seeds of life themselves! Living, moving beings: the essence of life . . .' and there have been many hasty judgments made like that in the recent history of science. But Brown first wished to check whether the same phenomenon was absent from similar-sized particles of an inanimate nature, from purely mineral sources. He turned his lens on powdered glass, fine-ground coal dust, metal powder and tiny mineral particles suspended in water – and found that they all showed the same effect. It was in this methodical and controlled manner that Brown analysed the occurrence of the unceasing microscopical movement that to this day bears his name.

The paper in which the account occurs was published in 1828 as a privately printed pamphlet with the title '*A Brief Account of Microscopical Observations Made in the Months of June, July and August 1827, on the Particles contained in the Pollen of Plants; and on the General Existence of Active Molecules in Organic and Inorganic Bodies*'. His account systematically describes the work he had undertaken, starting with the observations he had made on the *Clarkia* pollen and moving on through other species. Once in a while Brown's notes seem to hint towards a 'vitalistic' interpretation of events, which may echo the initial impressions he formed when his notes were first written up:

> I have examined the pollen of several flowers which have been immersed in weak spirit for about eleven months . . . and in all these plants the peculiar particles of pollen, which are oval or short oblong, though somewhat reduced in number, retain their form perfectly, and exhibit evident motion, though I think not so vivid as in those belonging to the living plant.

His account included the moment at which Brown realised that the active 'molecules' were not confined to the male sex cells in which he had first observed them with his simple microscope:

> The very unexpected fact of seeming vitality retained by these minute particles so long after the death of the plant

would not perhaps have lessened my confidence in the supposed peculiarity. But . . . I also found that on bruising first the floral leaves of Mosses, and then all other parts of those plants, that I readily obtained similar particles, not in equal quantity indeed, but equally in motion. My supposed test of the male organ was therefore necessarily abandoned.

Robert Brown went on to examine 'various animal and vegetable tissues, whether living or dead'; and found that the particles could always be released by bruising the tissues. When he moved on to pit-coal and fossil wood in oolite he pointed out that 'as I found these molecules abundantly . . . I supposed that their existence, though in smaller quantity, might be ascertained in mineralized vegetable remains.' So he moved on to non-living material, and furthermore, minerals that had no connection with long-dead organisms including all the minerals that Brown could lay his hands on – even 'a fragment of the Sphinx'! He listed only representative examples of the countless types of mineral to which he applied the test, and even that abbreviated list included

> travertine, stalactites, lava, obsidian, pumice, volcanic ashes, and meteorites from various localities, sand-tubes formed by lightning from Drig in Cumberland . . . manganese, nickel, plumbago, bismuth, antimony and arsenic. In a word, in every mineral which I could reduce to a powder, sufficiently fine to be temporarily suspended in water, I found these molecules more or less copiously; and in some cases, more particularly in siliceous crystals, the whole body submitted to examination appeared to be composed of them.

As the work went on, Brown showed his experiments to many of the callers at his busy home in Soho Square. What a range of people they were! Though Brown was a private man, an eccentric almost, and had a manner described as 'somewhat feminine perhaps, but never effeminate' (he never married) and seemed to have a terse and businesslike manner which was not given to the making of friendships, the quality and range of his work meant that Banks's former home remained the centre of British botanical science. During the summer of 1827 he demonstrated the phenomenon of Brownian movement to many of his acquaintances, all distinguished individuals. There were, he wrote, 'Messrs Bauer and Bicheno'. Bauer we have encountered as the artist who illustrated Everard Home's accounts of his investigations based on the Hunterian manuscripts whilst James Bicheno was at the time Secretary of the Linnean Society. George Bentham's letters of 1827 mention several visits he made to the Soho Square house, and in one of them he describes how Brown showed him the 'active molecules' under his simple microscope. Bentham says he asked for an explanation, but Brown dismissed the enquiry with the suggestion that all would be revealed in due course. Brown spoke similarly to Charles Darwin on a later occasion (p. 162).

Two other names Brown notes as: 'Sir Everard Home and Captain Home'. Home the elder certainly had an interest in simple microscopes and their performance, for he seems to have been responsible for depriving posterity of the Royal Society's bequest of Leeuwenhoek's silver microscopes (p. 69). His son was Sir James Everard Home who became a Captain in the Royal Navy. One of the visitors was an American by birth, Thomas Horsfield, born in Bethlehem, Pennsylvania, and a much-travelled naturalist and explorer. He worked under Stamford Raffles, the Singapore pioneer, as a naturalist and, on settling permanently in Britain, became keeper of the East India Company's museum in London until his death in 1859. The botanical collector and explorer Archibald Menzies appears in Brown's list, and he – though a trained surgeon – was taught by the same Professor John Hope in Edinburgh under whom William Withering had studied. Menzies became fascinated by plants and collected rare species from many parts of the world. He had been elected a fellow of the Linnean Society in 1790 and eventually became the society's father on the death of A. B. Lambert in 1842, then the only surviving original fellow.

Then there was Peter Mark Roget, to whom the Brownian movement demonstrations were shown; he was best known as the compiler of the *Thesaurus* which bears his name. Like so many of Brown's friends, he had studied at Edinburgh where he qualified in medicine. He was elected to fellowship of the Royal Society for his invention of a slide rule, and edited the *Proceed-*

ings of both the Society and its Council from 1827 until his retirement. Roget's obsessive tabulation of words under shades of meaning must have been a vast undertaking, and of course the work – suitably updated and revised – has been constantly in print ever since.

The final spectator to the phenomena whom we shall mention was William Wollaston, physician, scientist and botanist, a man who published a huge number of original observations and became known as a peerless innovator. A fellow of the Royal Society at 28, and of the Royal College of Surgeons at 32, Wollaston was not temperamentally cut out for medicine. The fate of his patients so stressed him that he felt he had to give it up: he wrote that 'the mental flagellation called anxiety' was such a burden as to make the loss of thousands of pounds 'a fleabite' by comparison. He invented optical devices, including for instance the Wollaston lens that was never produced in his life-time, but became popular thereafter; he also invented the *camera lucida*, which enabled an image to be projected on to a page so that it could be accurately drawn – and which in turn stimulated the development of photography. Wollaston was an admirable man, and doubtless advised Brown on botanical, microscopical and practical matters.

Here we have a pocket sketch, a pen-portrait, of this group of friends who met and talked, exchanged views and opinions, in the summer of 1827. As the weeks went by Brown put his notes on the 'animated molecules' in order and had them privately printed by Richard Taylor, of Red Lion Court, Fleet Street; the title page bears the words *Not Published*. Within a year he had to print a further pamphlet to set right some misconceptions, at least one of them resulting from the frankness of his initial exposition. Apparently there were some hasty appraisals by readers of his paper which did not follow the argument through, but with journalistic zeal latched on to the idea of 'active molecules' in the male germinal cell and misrepresented Brown's conclusions. He wrote:

> I have to notice an erroneous assertion of more than one writer, namely, that I have stated the active Molecules to be animated. This mistake has probably arisen from my having communicated the facts in the same order in which they occurred, accompanied by the views which presented

themselves in the differing stages of the investigation . . . I have formerly stated my belief that these motions of the particles neither arise from currents in the fluid containing them, nor depended upon that intestine motion which may be supposed to accompany its evaporation.

He continued by listing all those who might have made the same observation before him – Leeuwenhoek, Stephen Gray; Needham and Buffon (who argued between themselves across the English Channel about the question of spontaneous generation), Spallanzani, Gleichen (who first noted the presence of moving particles in pollen grains), Wrisberg and Muller, and James Drummond of Belfast. Brown was assiduous in affording the honour of precedent to those whose work had gone before.

Robert Brown did adopt one convention that is found in all branches of science – the reference to some side-issue in a paper ostensibly on something different. We saw how Robert Hooke used *Micrographia* as a vehicle for his thoughts on a wide variety of topics, and thereby put in print a host of novel notions not ripe enough to warrant a paper of their own. Brown did the same, and in one of his 'lateral digressions' gave us a term that is fundamental to modern biology – the cell nucleus.

The historic passage introducing the term comes in his 1832 paper on 'Fecundation in Orchideae'. Brown had undertaken a vast amount of methodical microscopy using his simple microscope, and he had noticed the presence of what we now know as a nucleus in the cells he examined:

In each cell of the epidermis of a great part of this family, especially of those with membranous leaves, a single circular areola, generally somewhat more opaque than the membrane of the cell, is observable . . . only one areola belongs to each cell. This areola, or nucleus of the cell as perhaps it might be termed, is not confined to the epidermis, being also found . . . in the parenchyma or internal cells of the tissue. The nucleus of the cell is not confined to Orchideae but is equally manifest in many other Monocotyledenous families; and I have even found it, hitherto however in very few cases, in the epidermis of Dicotyledenous plants.

Two decades later he was still publishing new microscopical ideas. In 1831 his pamphlet on 'Observations on the Organs and Mode of Fecundation in Orchideae and Asclepiadeae' was privately printed though later reprinted in the 1833 *Transactions of the Linnean Society*, and this most elegantly constructed example of pioneering microscopy describes how the pollen grains in these plants germinate on the stigma of the flower, and the pollen tube grows down to the ovary. This is the flowering plant's equivalent of sexual reproduction in the animal species with which most people are more familiar. The pollen grain is in some ways rather like the sperm cell, and on the surface of the pistil (the receptive extension of the female organs in the flower) the grain germinates, and the long, thin pollen tube grows down through the tissues of the pistil until it meets the ovary. There the male nucleus is able to fuse with the nucleus of the female egg-cell, forming the zygote. It is this, the first cell of the next generation, from which the embryonic new seed plant will form. The details are different from reproduction in animals, mainly because animals manage to maintain a moist pathway in which naked sperm cells can swim, whereas in flowering plants the pollen typically makes its way from stamen to pistil in a dry environment, and so it needs its resistant pollen grain 'coat' for protection and can only germinate into a naked, moist cell (the pollen tube) after it has reached its target. But – details apart – it is clear how the sexuality of flowers mirrors that of ourselves at the cellular level.

All this Brown documented in detail. Yet the pollen tubes themselves are tiny structures, no greater in diameter than a blood cell, and they are almost invisibly translucent. It is often relatively easy to watch pollen grains germinating on the pistil, but tracing the pollen tube down through the tissues to the ovary itself is a daunting undertaking. Yet Brown's work was done without modern conveniences, and (perhaps more limiting) without present-day insights to prepare the mind for what discoveries lay in store.

A third discovery Brown made with his single lenses was the circulation and streaming in the cytoplasm within the giant cells of the purple garden plant *Tradescantia virginiana*. These cells are beautiful to observe: strands of cytoplasm are woven across the cells, and tiny particles can be seen moving along in ordered rows like distant traffic on a highway. Sometimes they move at high speeds, and lines of movement in opposing directions occur adjacent to each other like a busy suburban road. Here Brown was watching the ceaseless activity of life itself, though he was perhaps disinclined to make too much of that. To this day, the cytoplasmic streaming in the giant cells of *Tradescantia* stamen-hairs, or in the large cells of the glassworts found in fresh water, is a memorable and instructive sight.

The important point about this work is that these sights are only possible with good microscopes. Even to see the cell nucleus in the crowded tissues of plants like orchids requires a good lens; but to observe particles small enough to indulge in vigorous Brownian movement, or the translucent specks of matter circulating in the stamen-hair cells, requires an excellent high-power lens. Brown's work was reviewed on the centenary of the naming of the cell nucleus, held in 1931 at the Linnean Society of London. The papers included a biographical summary of Brown's work by J. Ramsbottom, a talk on his zoological work aboard the *Investigator* by N. B. Kinnear, and an examination of Brown's activities with the Linnean Society by the then secretary S. Savage. The surprising fact is that there was no detailed comment or examination of Brown's microscopical methods, though doubts were raised about the ability of single-lensed microscopes to see half of what Brown claimed to observe. What compounds the omission still further is the astonishing coincidence that Brown's microscope had been found in a cupboard and returned to the Linnean Society virtually intact just a few years earlier!

10 | Bancks and the *Beagle* Voyage

The fate of Robert Brown's microscope could be easily followed from the documents in a small brown-paper envelope that was produced for me from the files by Elizabeth Young, then the Linnean Society executive secretary. Inside lay a black-bordered hand-written note which bore a strong and clear hand:

17 Dean Street
Feb[y] 5th 1859

My dear Bell,

I have been looking round for some trifling memorial of our late dear friend Rbt. Brown which might be acceptable to you, and it has struck me that the microscope (the common 'simple' microscope) which he was in the daily habit of using at the Museum, would possess greater interest in your eyes than any other object less immediately connected with those pursuits which made him what he was. With this feeling I beg your acceptance of it, without any apology for its little intrinsic value, simply as a relic; and I send it in the state in which it was left, without any attempt at cleaning, inasmuch as to purge it of its portion of its *erugos* would be to detract from its identity.

You are well aware that our great and good friend came of Jacobite parentage, his grandfather having been "out in the forty-five", and killed in the following year at Culloden. Now as I believe you also have a lingering regard for some at least of the Stuart race, I venture to send by the bearer what appears to me rather a curious genealogical tree of that unhappy family, and two or three old framed engravings which I suspect to have been heirlooms of loyalty to a hopeless cause. I regret that these are in so dirty a condition, but I send them merely as being possibly curious, and beg that if you do not think them worth keeping, you will at once destroy them.

    Ever your's faithfully
      John J. Bennett

John Bennett had remained Brown's assistant from 1827 until Brown's death in 1858. He had been a kind and methodical man, a school-fellow of John Keats the poet and John Reeves the actor (whose mathematics exercises Bennett used to do, in return for Reeves acting as his personal bodyguard in school fights). After Brown's death, Bennett had to defend the botanical collection from transfer to Kew, where W. J. Hooker and George Bentham had established their herbarium as an influential centre of botanical research. Official enquiries eventually established that the collections should remain where they were, in accordance with Brown's and Banks's wishes, but the strain on Bennett was considerable and he retired to Maresfield, Sussex, where he died of heart disease in 1876. Not all of Robert Brown's affairs were as easily disposed of as the little microscope.

The recipient of Brown's microscope was actually Thomas Bell, surgeon and naturalist, who was President of the Linnean Society between 1853 and 1861. He had been elected F.R.S. in 1828 and was one of the Society's secretaries from 1848 until 1853. Bell was a pioneer of dental surgery, but devoted most of his time to natural history and his *History of British Quadrupeds* (published in 1837) was reprinted in a revised edition in 1874, six years before his death at Selbourne. At a sale of his possessions the microscope was bought by a private collector whose daughter offered it to the Linnean

Society as a gift in 1922. Her letter, date-stamped on receipt by the society 19 January 1922, says:

> Amberley
> Reigate
> Jan. 19. 1922
>
> Dear Sir
>
> By the kindness of Mr Salmon, I have much pleasure in offering Mr Brown's microscope to the Linnean Society if they care to accept it. Its credentials are in the box with it. At the sale of Mr Bell of Selbourne's effects, it was bought by my father & so its history since its original owner is accounted for.
>
> Yours faithfully,
> Ida M. Silver (Miss)

The microscope itself was lying in its original mahogany box, in a dirty and blackened condition and showing signs of actual damage – parts of it were bent and buckled, and some of the components were misplaced, lying in the wrong places and apparently wrongly identified (the screw cap that belonged to the top of the microscope pillar, for instance, had been put into one of the holders designed for the lenses as though its purpose had been mistaken). But it was a beautifully designed portable microscope. The circular stage slotted into a dovetail bracket attached to the main brass pillar, and there was a concave mirror – by now badly marked – which fitted beneath the stage. Two of the lenses were fitted with Lieberkühn reflectors which were dull grey and pitted-looking. Inside the case was glued a small piece of paper on which was written:

> This microscope belonged to Robert Brown, and was certainly used by him for many years, until within a short time of his death, which event occurred on the 10th June 1858, in the 85th year of his age. The instrument was given to me, as a memorial of my revered old friend, by his Executor, my friend John Joseph Bennett Sec. L. S. on the 5th day of Feby. 1859. Thomas Bell.

The microscope seems to have been neglected until the 1970s, when the Linnean Society secretary at the time – T. O'Grady –

and one of the fellows, W. A. S. Burnett, examined it and Burnett took a photograph of the microscope in the assembled position. Professor Irene Manton F.R.S. also showed interest in the little microscope at the time, and her colleagues showed it was possible to obtain photographs of onion epidermis and *Tradescantia* cells using the low-power lenses.

The microscope had narrowly missed one brush with fame when the great 1951 Festival of Britain organisers asked to show Brown's microscope in the Dome of Discovery. The Society's Council at the time turned the request down, because of the precious nature of the instrument. It seems probable that one reason for declining the suggestion was that the microscope was so very dirty, neglected and damaged that it would have looked decidedly out of place amidst so many more eye-catching exhibits. Early in 1981 the Council agreed that I might remove the microscope for critical study and renovation. The project was to examine the instrument, to realign the damaged components, calibrate the lenses, and put it back into something like the condition it was in when Brown used it. Here was a wonderful opportunity to find out something about the way these single-lensed microscopes were used in practice, and to make a series of experiments that would repeat Brown's original discoveries, and show once and for all just how good such a microscope was in practice – not on paper, not in theory; but on the laboratory bench. The problem was the delicate one of conserving the authenticity of the instrument whilst returning it to a usable condition: the conflict between Bennett's view in his letter of 1859 that to remove 'any portion of its *erugos* would purge it of its identity' and the words of Professor Manton, in her George Bidder Lecture at Leeds University in 1974, that 'its condition . . . is not very good since minor repairs are needed'.

If ever there was a euphemism, here was one! The term *erugos* is itself a fascinating one to interpret. There is a word *erugate* which means 'to remove wrinkles from' and is derived from the Latin verb *erugare* which itself has roots in *ruga*, wrinkle. Manton translates the term as 'wrinkles' in her account. My view is that the term was *erugo*, or, more usually, *aerugo*, which connotes the verdigris that forms on brass and actually derives from the Latin *aes, aeris,* = brass. In this sense the term is closer to what Bennett must have meant, for the connotations of 'aerugo' have the same sense of quality and authenticity as *patina*, which describes both the incrustation on old bronze and the bloom on antique furniture. I am sure that is what Bennett implied – there is no such word as 'erugos', and I expect that was just a slip of his pen. But as the contemporary accounts of Brown's life suggest, he lived as the years passed in increasing disarray and with little cleanliness in his surroundings, a fact substantiated by Bennett's description of the other relics he gave to Bell – all of which seem to have disappeared since. From this it is clear that Brown's microscope was, simply, in dirty condition and no amount of euphemism can disguise that. In addition it spent over sixty years in the hands of Bell and, after he died, of Silver before being returned to the Linnean Society, and there is no knowing how much damage it sustained during that time.

I therefore decided to remove the superficial dirt from the microscope, to examine and document it thoroughly for the first time in its history, and to make good the damage that could be repaired without the need for outside components. As an example, there was an unsightly hole in the padding of the lid of the case which I repaired by fitting a matching piece of fabric behind the gap, but without securing it in place. The effect is that the case looks nearer its original condition, but the damage remains demonstrable should anyone else wish to inspect it in the future. By adopting this approach it was thought possible to maintain a correct balance between the need to protect and refurbish the microscope on the one hand, and the desire to keep it in its original condition on the other.

The microscope was indeed in a sad state. The brasswork was pitted and blackened, the stage was partly separated from its focusing collar and someone had apparently made an ill-advised attempt to straighten it by applying pressure to the whole assembly, as its bracket was bent downwards. The dovetail into which the stage should fit was splayed out and loose, the cap that tops the main pillar was misplaced and damaged, the teeth on the main focusing pinion were worn, and as well as a portion missing from the lining of the lid, the seal at the base of the mounting bush on the outside of the lid (into which the microscope screwed during use) was also missing. The box itself was marked and damaged, one of the securing hooks for the lid was broken in half, and the lid lining pad was the wrong way round. The lenses were dirty and in some cases wrongly assembled, and even the ribbon loops used to lift out the case liner – into which the components were custom-fitted – were broken.

In restoring the instrument so that it could be used again, none of the signs of 'wear and tear' were interfered with. Missing pinion teeth were not replaced, nor was the mirror resilvered, for example. The aim was to make the microscope whole again, by making good the accidental damage it had sustained since it ceased to be a useful instrument of scientific investigation. And what a delight the microscope turned out to be in use! There were six lenses, ranging in magnification from ×5 to ×170 and providing crisp and sharp images. Plant tissues stood out beautifully; such smaller-than-average cells as human red blood corpuscles and yeast cells showed with clarity; Brown's nuclei were as plain as a pikestaff and so were bacteria. Two of the lenses were fitted with Lieberkühns and – though they looked in need of complete resilvering – it was found possible to polish them again and use them to observe opaque specimens. The effect of this form of illumination was beautiful: colours, structural detail, shapes and outlines were graphically revealed by the all-round illumination of the concave mirror. Part of the value of the Lieberkühn is that the light source shines through the translucent portions of a specimen in addition to reflecting light on to the surface nearest to the observer. Insects, for instance, show up in a startlingly revealing fashion unlike anything you would find on a conventional modern instrument.

The microscope itself was photographed in detail, and it is plain how it works. The lens-holder fits at the end of the lens bracket, and a knurled knob allows one to rack the lens across the specimen as it lies on the stage. There are two means of focusing the lens: a rack-and-pinion focusing control fitted to the main pillar, and a cylinder of cork that is fitted into the upper end of the pillar. That enables the user to slide the lens up or down until the image is roughly focused, before using the focusing knob to adjust it in detail. I replaced the damaged original mirror with one in better condition for some of the pictures, and added some well-preserved accessories from other microscopes in order to obtain photographs of the instrument as it would have looked when it was in active use. These

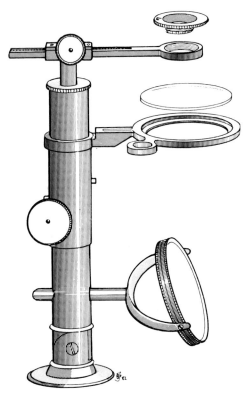

Fɪɢ. 24 Robert Brown's microscope at the Linnean Society. Visible are the two milled wheels controlling the focus and the lens arm. Note also the screw-in lens cup and the circular glass stage. The instrument could be inclined through a fulcrum at the base. It dates from before 1820, and produces clear images of good magnification.

views provide a unique impression of how Brown would have known the microscope in its early years – but they were temporary changes for the sake of the record, and the microscope today once again lacks these extra touches.

Reconstructing Brown's experiments was the highlight of this work. Orchid tissue was obtained from a major botanical collection in Wales, and I mounted thin portions on crude glass discs, without any extra mountant, coverslips, or other more modern aids, using nothing but the plant sap – released during preparation – as a mountant. The nuclei were well revealed, and under the higher-power lenses minute cell structures stood out clearly.

I turned his lenses on the giant cells from the stamen hairs

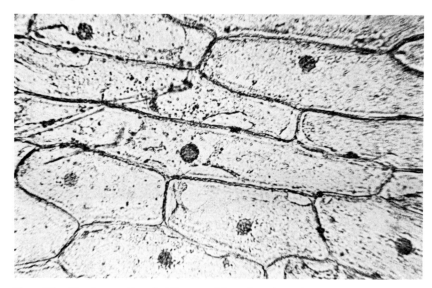

FIG. 25 Twelve cells of *Allium* epidermis, viewed through a medium-power lens of Brown's microscope (Fig. 24), show cell-walls and nuclei with great clarity. Contrary to popular belief, aberrations are slight, and the image quality compares favourably with modern microscopes fitted with fully corrected lenses.

of *Tradescantia*, again using unsophisticated techniques of microscopy. The lilac-coloured cells stood out beautifully, and under the higher powers there were the streaming cytoplasmic strands, the tiny particles gliding along like purposeful entities with an aim in view. Suspensions of particles in water (I used Indian ink for this experiment) showed how clearly the lenses revealed Brownian movement, which compared well with observations I made using the excellent NPL lenses of our modern Leitz microscope, although the modern Leitz images are, of course, better.

Brown's microscope was different in many respects from the early aquatic microscope – the so-called 'Ellis' microscope. The stage is signed with the maker's name – BANKS 441 STRAND LONDON. Robert Banks (whose name was usually spelt Bancks, and for consistency – and to avoid confusion with Sir Joseph Banks, who was in no way related – I will keep to this less familiar spelling) flourished from 1796 as an instrument-maker. From 1820 he was joined by a son. Bancks – and, later, Bancks & Son – were instrument-makers who produced simple and compound microscopes. An example of a compound instrument dated 1815 is in the Royal Microscopical Society's description of its collections published in 1928, and it is signed: *Banks, Inst. Maker to the Prince Regent, 441, Strand, London.* Robert, the father, was in the Strand until he was joined by his son. An engraving made some years after Bancks & Son had moved to an address in New Bond Street shows the old building well. The scene is little altered today. Behind the facade of the old Regency terraces is an ultra-new suite of offices with air-conditioned terraces and waterfalls playing amongst potted trees and palms of many kinds. But to the visitor who emerges from London's Charing Cross station in the Strand, the cream-painted buildings across the street look to have changed but little since the time when Robert Bancks made microscopes for the Prince Regent.

Robert Brown's own accounts show that he must have had a powerful microscope when he investigated pollen before 1810. A reasonable assumption is that he used this instrument for that work – after all, a less powerful microscope would simply not have given the insight needed to draw the conclusions he did. The address on the microscope shows us it must have been made during the period that Bancks was based in the Strand – i.e. between 1796 and 1820 – and the date of Robert Brown's work (1810) fits that well.

Later in his life, Brown used other instruments. His account of Brownian movement (as we now know it) begins with a note:

> The observations, of which it is my object to give a summary in the following pages, have all been made with a simple microscope, and indeed with one and the same lens . . . which I obtained from Mr. Bancks, optician, in the Strand. After I had made considerable progress in the inquiry, I explained the nature of my subject to Mr. Dollond, who obligingly made for me a simple pocket microscope, having very fine adjustment, and furnished with excellent lenses, two of which are of much higher power than that above mentioned. To these I have often had recourse, and with great advantage, in investigating several minute points. But to give greater consistency to my statements, and to bring the subject as much as possible within the reach of general observation, I con-

FIG. 26 The microscope made by Dollond of London of the type Brown recommended. The highest-power lens magnified ×480, but has been missing for some years. It was one of the most powerful lenses ever to be ground. The focusing control is adjacent to the stage itself, and the whole microscope is produced in lacquered brass to a high degree of precision.

tinued to employ throughout the whole of the inquiry the same lens with which it was commenced.

The focal length of the ×170 lens – on the traditional ten-inch standard – would be $\frac{1}{17}$ inch, but Brown says his lens had a focal length of $\frac{1}{32}$ inch. That is close to the *working distance* of the lens (i.e. the gap between the front of the lens and the specimen when it is properly in focus), and I believe that Brown was actually referring to the ×170 lens in his account. It has been claimed that the Dollond microscope was now preserved in the Kew Museum. The address at the opening of the International Meeting on Botanical Microscopy at York in July 1979 concluded that: 'The microscope in the Kew Museum agrees with the description given by Brown of his Dollond instrument. It has lenses of diverse sorts, and has a fine-focusing adjustment.' However, when one inspects the microscope at Kew it is clear that it *cannot* be the one to which Brown referred. In the first place, by no stretch of the imagination is it a 'pocket' microscope. It is much the same size as his earlier instrument. Secondly, though it does have various lenses, none of them is of the higher magnification he describes – and two of the lenses with the pocket microscope were stated to have a power that was 'much higher'. Third, the microscope was clearly not made by Dollond, but by Bancks – it says so in engraved letters round the stage.

The answer to the puzzle lies in an entry in the catalogue to the University Museum of Utrecht, which was drawn up in 1934 by P. H. van Cittert. On page 28 appears the heading: *Pocket Microscope of Robert Brown.*

> This instrument bears the signature '*Dollond London*'. It was bought by the Physical Laboratory for one hundred guilders early in the nineteenth century. Its construction differs greatly from that of the ordinary Cuff type. The lenses . . . magnify ×185, ×330 and ×480. Harting gives for the missing lens a power of ×77 . . . he gives the powers of the lenses and comments upon the ×480 lens as being the strongest lens obtained by grinding and polishing he had ever seen.

The matter does not end there, though, for the catalogue is not faithful to the original entries in the accessions register of the University Museum. Entry No. 419 shows a handwritten description of the microscope in a red leather morocco case and then says *volgens Rob. Brown*. Literally, that reads *following Robert Brown*, which could mean 'after his design' or even 'made to his prescription'. But it does not simply say 'Brown's'. To my mind that introduces some uncertainty. It is unlikely that

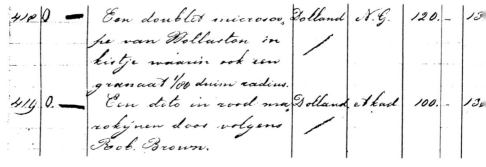

FIG. 27 The microscope in Fig. 26 is widely quoted as being 'Robert Brown's', though this entry in the accessions book of the Utrecht University Museum shows clearly that it 'followed' or 'was after' Brown – the words are 'volgens Rob. Brown.' – as discussed in the text.

Brown, if he had taken possession of a new pocket microscope made specially by Dollond, would have sold it to the Physical Laboratory in Utrecht 'early in the nineteenth century'.

On the other hand, two of the lenses certainly do have magnifications much greater than those of his earlier instrument ($\times 330$ and $\times 480$, compared with $\times 170$). In summary, the microscope at Utrecht most certainly fits Brown's description insofar as it *does* have a 'very fine adjustment', the lenses are 'fine' and two of them are 'of much higher power'; and the instrument is certainly a 'pocket microscope'. Finally, it is signed by Dollond. It may not be that this is the same instrument that Brown used, and to which he referred in his paper. But it is certainly the same design exactly, and even if it might not be the same individual microscope he used for his Brownian movement research it is clearly that instrument's twin. Dollond may well have made virtually identical copies of the design for other users.

Where does this leave the Kew microscope? It certainly belonged to Brown: set into the lid is a silver plaque which reads:

The Microscope used for years
BY ROBT. BROWN F.R.S.
in Soho Square
Given to me by J J Bennett F. R. S.
12 Feby 1859   *J. E. Gray*

It seems as though Bell was not the only associate of Robert Brown who received the favour of one of his personal microscopes. In this case the recipient was John Edward Gray (1800–1875), the naturalist son of Samuel Gray (1766–1828) the younger, in his time a noted naturalist. Both father and son used to lecture on Jussieu's system of classification, and the son – together with the assistance of such men as de Candolle and Bennett – was really behind the influential *Natural Arrangement of British Plants* (1821) that appeared under the father's name. J. E. Gray was a controversial figure with strong likes and dislikes. Robert Brown certainly did not care for him, at least at

FIG. 28   The simple microscope in its final years, exemplified by this fine example in the collections at the Herbarium, Kew. It was Robert Brown's property, and features a rotatable double-sided mirror and a fine-focusing control mounted below the coarse-focusing knob. As a research instrument for student or field use, it would be well fitted for present-day applications (p. 156).

first, and indeed Gray was actually blackballed when he applied for election as a fellow of the Linnean Society because of an assumed slight against the President, Sir J. E. Smith. After his rejection, he turned his attention to zoology and at the British Museum from 1846 to 1875 was a colleague of Brown and Bennett. He was an ardent campaigner and innovator; Gray always insisted that what eventually became the Penny Black had originally been proposed by him.

The microscope that Bennett passed on to Gray is a superbly refined version of Bancks's earlier design. The mirror is two-sided, and as well as being held in a swing pivot so that it can be rotated about its own axis, it is fitted into a rotating collar so that it can be turned through 360° around the main pillar. The lens arm has an inset pivot so the lens itself can be twisted around the complete circle if need be. At the base of the main pillar is a fine-adjustment control in the form of a double-milled wheel that can be rotated by a hand resting on the case into which the microscope fits during use. The whole instrument is cleverly designed and is already beginning to show the trend towards a microscope produced for the gadget-conscious microscopist. When I took the fine-adjustment apart for examination, a screw thread was seen in the base of the pillar which does not play a part in the mechanism, but seems to be related to an alternative design for this fine-adjustment. This is a useful reminder that in those days before mass-production, microscopes had an identity of their own. No two were alike: the tendency was for production on a one-off basis to a precise order from the customer, and even when they were made in advance they were not always exactly similar. For instance, parts are not interchangeable on the close-fitting components of Bancks's microscopes. This can be related to the conclusion that because the long screws in two of the silver Leeuwenhoek microscopes are interchangeable, this proves beyond doubt that they were 'twins'.

And what of the lenses to these microscopes? The microscope at Kew has a range of magnifications from ×12 up to ×160 – nothing quite as powerful as the earlier instrument, true, though there are some strange little lenses which appear to be doublets. That is, they are made with two lenses fitted together, rather than just one. This was an early attempt to improve the optical performance of a microscope, but the experiment was not a success. In each case these lenses are inferior to the simple lenses. A couple of the doublets have such a short working distance that they seem to have actually grounded on the specimen, and the surface of the lens is rough and scratched. The optical performance of the Dollond lenses with the Utrecht microscope is very good; here the lenses are fitted into little sliding holders, rather than round holders as was usual with the Bancks type instrument. Much to my regret the highest power lens is now missing and has been so for some years by all accounts. It cannot have been much larger than 1 mm in diameter, the size of a grain of sand or a printed full stop, and I had some hopes that it might have dropped from its holder into a recess within the felt-lined case. Some specks of dust were there, right enough, but there is no sign of the lens. We must be grateful to van Cittert for documenting it when he did, for Dollond's precision in making this lens was formidable. The image would have been hard to adjust, not very bright, and extraordinarily hard to observe; and it may be that such a high magnification, which puts the lens into a class of its own, was beyond the practical limits of manipulation.

A chance conversation with Gavin Bridson, then the librarian at the Linnean Society, led me to two other microscopes of the same type, also made by Bancks, and like the Brown microscope described above also at Kew. Gren Lucas, deputy keeper of the Herbarium at Kew, button-holed me at one of the Linnean Society's receptions and wondered if I would be interested in a couple of early microscopes that were part of the Kew collection. The instruments represent two important examples of Bancks simple microscopes. One of them had belonged to George Bentham, the other to Sir William Hooker, a long-standing friend of Brown's. Bentham was an acquaintance of Brown's, and he had been present at the demonstrations of Brownian movement in 1827 (though Brown did not count Bentham amongst those names worthy of record in his own account); whilst Hooker was the Director of Kew from 1841 until his death in 1865 when his son, Joseph Dalton Hooker, succeeded him in that office. His microscope still has seven lenses in the case, each carefully fitted to a purpose-built holder, and ranging in magnification from ×8.5 to ×135. I found each of them capable of producing a good image, five of them excellent. An engraved trade label is glued inside the lid, giving the name of the manufacturer as 'BANKS' and the address 441 Strand, London, as before. The microscope has with it a small typed label reading:

> MICROSCOPE used by Sir W. Hooker, and
> employed by him in describing the 'BRITISH
> JUNGERMANNIAE', 'MUSCI EXOTICI' &c.
>   Sir J.D. Hooker, G.C.S.I. &c.

Further evidence links the instrument with William Hooker, and I found it on the label that accompanies the key to the case. On the reverse are the letters '. . . m H . . .' for the label is cut from the central portion of an old and brittle visiting card. It provides an interesting confirmation of the microscope's provenance.

Bentham's microscope is a beautiful instrument in excellent condition. Provenance lies in the fact that his name is written inside the case – apparently in Bentham's own hand – whilst the details on the label that now accompanies the instrument read thus:

> DISSECTING MICROSCOPE
> Used by the late George
> Bentham, Esq., F.R.S.
> Presented by Sir Joseph Hooker, K.C.S.I.
> 37.   1890.

The idea that these simple microscopes were *dissecting* microscopes became a popular misconception. This does not mean that the microscopes were excluded from dissection work, for they were used to help minute examination of material like any microscope of our own time. The point is that a dissecting microscope is a particular type of instrument, a low-power magnifier with a limited purpose. The superficial similarity between aquatic microscopes and modern dissecting instruments led to the term being used indiscriminately.

I can demonstrate that Bentham used his microscope for dissection, without doubt. Firstly, of the two lenses found with the microscope, one is a fine low-power lens magnifying $\times 19.6$ times. This lens is ideal for use when dissecting. The use of a modern microscope has revealed to me the signs of his work

Bentham left on the microscope, for there are many tiny incisions along the edges of the lens bracket of his microscope, which my own lenses reveal. They were made by a sharp steel blade, and show where Bentham used the top of his microscope as a tiny dissecting block. From this 'forensic' datum one can conclude that the microscope may have been used in the field, for on a laboratory bench or in a study there would be a host of other steady surfaces on which to dissect a specimen . . . and that, as I will explain in a moment, helps us fix a date for this instrument! The other lens with the microscope would be of no value at all for dissection, since it magnifies ×170 and is capable of generating superbly crisp and well-defined images. So – though Bentham's microscope was capable of being used as a dissection aid, and shows signs that it was undeniably used in this way, it was not merely a dissecting microscope in the accepted use of the term. This was, as were all the Bancks microscopes, a well-made research microscope capable of revealing details that would be associated with a high-quality modern research instrument.

How can we date these microscopes? One valuable aid is the historical record of the British Royal family, for Bancks was a manufacturer of optical instruments for the Prince of Wales, later George IV. We have seen how he styled himself *Instrument Maker to the Prince Regent* on a microscope made in 1815, and the lack of such a designation on the Linnean Society microscope suggests it dates from earlier than that. As I have shown, Brown would have required a microscope of exactly this type for his work on pollen published in 1810, and we might date the instrument from around that time. The fact that Hooker's microscope was used in his research for his book *British Jungermanniae* gives some indication of its date, for the book was published in 1816 and Hooker was working on it some years previously. However, the inscription reference to the 'Prince Regent' in the instrument in the Royal Microscopical Society's collection to which I referred on p. 151 contrasts with the inscription in the lid of Hooker's microscope, which mentioned 'His Royal Highness *The Prince of Wales*'. The Prince became Regent after much negotiation and controversy in February 1811, and so the microscope is likely to have been made prior to that date. Hooker began life, in fact, as an ornithologist and then an entomologist and it was around 1805 that his discovery of a rare moss encouraged Sir James Edward Smith to recommend that Hooker should take up botany. In 1806 he started botanical studies in the Scottish highlands in the company of Dawson Turner F.R.S., his future father-in-law, and Sir Joseph Banks sent him on an expedition to Iceland two years later. We might speculate that the microscope would have been needed for these trips, and if so it could be dated *circa* 1806. The appearance of the microscope lends support for it being older than Brown's Linnean Society instrument, for the Hooker microscope has earlier flanged lens holders for the low-power lenses, whilst the higher-power lenses are fitted with domed screw-in protecting caps similar to those found with late-eighteenth-century microscopes. The microscope case is more sparsely fitted – Bancks installed plush padding in all the other, later, microscopes – and the instrument is unique in the series in having a paper plate pasted into the lid, instead of the engraving on the stage. This is probably the oldest of them, and represents the type of instrument on which Bancks was experimenting with the design he was later to bring to a greater degree of sophistication.

Bentham's microscope provides a clear date, for it is engraved with the words:

BANKS & SON. Strand London, Instrt. Makers to HIS MAJESTY

Now, it is known that Bancks left the Strand address in 1820 – it is recorded in that admirable compilation *English Barometers 1680–1860, A History of Domestic Barometers and Their Makers* by Nicholas Goodison, published by Cassell, London in 1969. In 1820 the Prince Regent ascended the throne as King George IV, and so the coincidence of the Strand address and the reference to 'His Majesty' means that the microscope must have been made in 1820. Next in the sequence would be the Dollond microscope Brown referred to in his paper on Brownian movement, for the circumstances he mentions in that publication imply that he obtained that microscope in 1827.

Following that instrument would come the microscope with the fine-adjustment which is now in the Kew collection, and which Brown possessed at Soho Square. The footnote to Brown's paper *q.v.* suggests that when he used the Dollond microscope he already owned *one* made by Bancks, and so we might speculate that at the time he did not own this more sophisticated microscope. In addition, he refers to the exquisite fine-focusing control on the Dollond microscope as one 'having very delicate adjustment', so it was clearly a novelty for him. This later Brown microscope *also* has a delicate adjustment, and the fact that it is not referred to in that publication substantiates the conclusion that in 1827 he did not own it. There is another microscope that allows us to put forward a date a year or two after this, as it happens: this is Darwin's microscope, which he took with him on the celebrated voyage of the *Beagle*. It is still in the Down House, where he lived. Unfortunately no lenses survive with the microscope – though since Darwin was friendly with Brown, Hooker and Bentham it is just possible that some of the lenses at Kew belong to this instrument. There is one somewhat ungainly simple microscope, made by Chevalier of Paris in the early nineteenth century, attributed to Sir William Hooker in the Kew collection, and the drawer in the case is crammed with contemporaneous instruments for handling specimens and even a tiny 2 mm-square microphotograph showing a portrait of two ethereally beautiful young women. There are many lenses present, too, and I do not imagine they all necessarily belonged to the microscope. Perhaps a few of them belonged to the Darwin microscope at Down House after all.

As the photograph (Fig. 29) shows, it is similar to the microscope of Brown's now at Kew – it features the same fine-focusing control at the base of the pillar, though with the earlier way of mounting the pivoted sub-stage mirror. That, and the condenser lens that is also found below the stage, relate it to the earlier Bentham microscope. On that basis one might argue that this was a slightly earlier microscope than the one owned by Robert Brown; Darwin's would be dated at around 1829, and Brown's a little later. On the other hand a strictly chronological sequence based on design sophistication is not necessarily reliable. Nowadays we are so used to the need to acquire the 'latest model' of anything from a shirt to an automobile, from a bunsen burner to microscopes for research, that it is hard to imagine an era when purely practical considerations mattered more. Instrument makers like Bancks produced what the customer required, and it is likely that the

Darwin microscope has fitments on it requested by Darwin, just as Brown would have asked for what suited *him*; and it might be more realistic to put these two microscopes – with their characteristic fine adjustment – into the 1829 bracket without stressing too much that one preceded the other. In either event, Brown advised Darwin to obtain this Bancks instrument for the *Beagle* expedition, which left England in 1831; and it could be that he already owned his example of this genre.

When Charles Darwin was an old man he wrote:

> Robert Brown . . . of whom I saw a good deal . . . seemed to me to be chiefly remarkable for the minuteness of his observations and their perfect accuracy. His knowledge was extraordinarily great and much died with him owing to his excessive fear of ever making a mistake. He poured out his knowledge to me in the most unreserved manner yet was strangely jealous on some points. I called on him two or three times before the voyage of the *Beagle* and on one occasion he asked me to look through a microscope and describe what I saw. This I did, and believe now that it was the marvellous currents of protoplasm in some vegetable cell. I then asked him what I had seen, but he answered me, 'That is my little secret!'

It is probable that Brown used the microscope now at the Linnean Society for his discovery of Brownian movement, supplemented by observations made with the kind of microscope at Utrecht; he may have used the same Linnean Society microscope with his later Bancks instrument (now at Kew) for his work on the nucleus, and probably used the Kew microscope for the work on the cytoplasmic streaming in the cells of *Tradescantia*.

Thus we have moved from the basic simple microscope made by Hooke and Leeuwenhoek up to these instruments made a little before 1830 by Robert Bancks of London – from the most unrefined, though effective, little microscopes imaginable right up to a finely designed instrument with every convenience the user might require – including a fine-adjustment for the focusing that is more than adequate, even if you are used to the functioning of a modern instrument. Many of the most far-reaching developments of microbiology and the study of microscopic structure were made with instruments such as these, and the universally accepted convention in museums and universities all round the globe to label them 'dissecting microscopes' – i.e. microscopes limited to crude, low-power work – *must* be extirpated if facts are to remain facts. For all but the finest detail, simple microscopes can be used for a huge range of research. Indeed it is true to say that almost every microscopical discovery of note could have been made with nothing more complex than a simple microscope of the Leeuwenhoek type – the clarity of our modern vision owes far more to improvements in the way we *prepare* specimens for microscopy than it does to the instruments themselves. Certainly the relatively obscure name of Bancks should be recognised as a major designer of microscopes, since he produced them for Darwin, Hooker, Bentham and Brown – as well as for the King himself. Not only that. The conventional view of the history of the microscope is that as compound instruments came to the fore, the design of these simple microscopes faded to extinction.

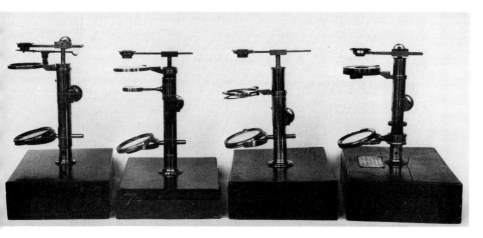

FIG. 29 Assembled for the first time (at least for 150 years), the microscopes owned by (L-R) Brown, Bentham, Hooker and Brown. Though they are of the type designated 'low-power dissecting microscopes' they are all capable of generating high magnifications and produce images of high quality. Note accessories such as the substage condenser lens or the racking control for the lens-holder. Each was built, according to the purchaser's specifications, during the first two or three decades of the nineteenth century.

I am sure that is equally erroneous. By the time Bancks had perfected his design, he had set in motion the trend that set design standards for the best compound instruments. In his construction lie the origins of today's most advanced microscopes. Not only is the simple microscope worthy of a better historical image, and even worth considering for resuscitation, in my view it should be seen as the true ancestor of the modern research instrument. When the simple microscope was superseded (which is the area explored in the next, final chapter of this book) it did not entirely disappear, for its design principles are with us today.

11 | In Consequence

The dawn of the modern compound microscope is heralded in
Robert Brown's writings. He had some difficulty in measuring
the size of the little particles he observed in Brownian move-
ment: at first he calculated that they were '$\frac{1}{15,000}$th to $\frac{1}{20,000}$th of
an inch', but then he added a note that 'Mr Dollond had
examined some spores 'with his compound achromatic microscope'. The experiment was done as his pages were being
passed for the press, i.e. in the early summer of 1828. Dollond's
conclusions were that the particles 'were about $\frac{1}{20,000}$, yet the
smallest did not exceed $\frac{1}{30,000}$ inch'. These dimensions are of the
order of 1 micrometer, and it is interesting to see that Dollond's
achromatic instrument could at that time provide evidence of
smaller particles than Brown could claim to resolve with his
own simple microscope. It was a prophetic observation – from
that time on, the achromatic experiments began to produce
results that were superior in quality to those of single lenses
and, though there was no palace revolution, the trend continued until the middle of the century. Around 1850 the
compound microscope – as generally available – had superseded its simple predecessor. Even then the simple microscope
was still employed. Darwin wrote a letter in March 1848 to

Richard Owen, the first president of the Microscopical Society
of London (which was founded in 1839, and became the Royal
Microscopical Society in 1866). In it he said:

> I dare say what I am going to write will be absolutely
> superfluous but I have derived such infinitely great
> advantage from my new simple microscope, in compari-
> son with the one I used on board the Beagle, and which
> was recommended to me by R. Brown, that I cannot forgo
> the mere chance of urging this on you.

He goes on to speak of that microscope as being thought of as
'the best' just a few years before – and this, remember, in 1848.
The final downward turn in the fortunes of the simple microscope can be glimpsed in the words of William Carpenter
M.D., F.R.S., writing in 1862:

> I suggested to the Society of Arts in the summer of 1854
> that it should endeavour to carry-out an object so strictly
> in accordance with the enlightened purposes which it is
> aiming to effect . . . to aim at obtaining . . . a *simple* and
> low-priced microscope for the use of Scholars, to whom it
> might be appropriately given as an award for zeal and
> proficiency in the pursuit of Natural History . . .
>
> *Field's Simple Microscope* consists of a tubular stem,
> about five inches high, the lower end of which screws
> firmly into the lid of the box wherein the instrument is
> packed when not in use. To the upper end of this stem the
> stage is firmly fixed; while the lower end carries a concave
> mirror. Within the tubular stem is a round pillar, having a
> rack cut into it, against which a pinion works that is
> turned by a milled-head; and the upper part of this pillar
> carries a horizontal arm which bears the lenses, so that, by
> turning the milled-head, the arm may be raised or lo-
> wered, and the requisite focal adjustment obtained. Those
> who have trained themselves in the application of it to the
> study of Nature, are well prepared for the advantageous
> use of the Compound Microscope.

The magnifications of the lenses supplied ranged 'from about
five to forty diameters'. This was the last period when the

simple microscope was regarded as a serious tool of study – and even here it was recommended as being 'most fitted for the *preparation*' of objects, with the compound microscope being reserved for their examination.

All the time that the simple microscope was slowly developing through the two centuries over which it held sway, the compound microscope was changing. From the several versions of the instrument supported on a tripod to those that were linked to a side support, the final form that emerged used a system of load distribution based on a triangle: the supporting arm rose from the level of the stage to carry the body tube – equipped with the lenses – and its focusing controls. The traditional concept of a microscope fits this model perfectly; almost every film and television presentation showing the peerless scientific investigator hunting for truth in his cloistered laboratory features one of these. It was this type of microscope that John Joseph Lister selected for his design of a fully achromatic instrument. Lister was no relation to other famous Listers; he had left school at fourteen to help his father in the wine trade and learned mathematics through spare-time study. It was in 1830 that he showed how to use achromatic doublet lenses in a microscope without (as was previously thought) producing severe spherical aberration. The details of all this do not concern us in a survey of the simple microscope – but his work did mean that, from then on, lens systems could for the first time be *designed*, rather than merely produced on the basis of practical trial. J. J. Lister gave us the modern system of optics, no small achievement for a man who never went to college.

The adoption of the triangular configuration for his instruments meant that they soon became known as *Lister type* microscopes. This kind of arrangement is difficult to manufacture, and the history of the compound microscope tells how in 1843 a different type appeared with a transverse bar or limb supporting the body tube. The design was announced almost simultaneously by the two greatest microscope manufacturers, Andrew Ross and the firm of Powell & Lealand. Ross was an instrument-maker who had begun his work around 1831 and had attracted a reputation for precision and craftsmanship. Hugh Powell made his name by producing the first microscope with an achromatic condenser mounted below the stage (the

condenser he imported from Europe) before going into part-
nership with his brother-in-law P. H. Lealand. Because of the
peculiarity of the design, this kind of microscope became
known as the *bar-limb type*. From 1843 it dominated the field of
high-quality, precision compound microscopes. Two Oxford
students of the history of the instrument record how important
this design was. Wrote Savile Bradbury in 1967: 'This bar-
limb model, as it became known, was adopted from Powell &
Lealand and was used by most English makers for some
considerable time.' And at about the same time (in a paper
dated 1966) G. L'E. Turner recorded that: 'This announcement was to be a momentous one, for the design of the
microscope described became the basis of the firm's instruments for over sixty years.' Indeed, catalogues of precision
microscopes have featured bar-limb microscopes since that
time, largely for the good reason that a system in which the
support for the body tube of the microscope is made to be
exactly parallel to the stage is more reliable than producing one
at an angle.

Turner wrote, in allusion to the simultaneous announcement of the same principle by two manufacturers: 'It is difficult
to say who was the inventor of this type of mounting.' Though
it may prove impossible to decide between Ross, and Powell &
Lealand, the student of the simple microscope will immediately recognise that the bar-limb is nothing more than the essence
of the microscope type produced by Bancks, among others.
Many of the types of microscope that had been produced in the
preceding decades had allowed one to unscrew the single
lens-holder from a bar-limb-type aquatic microscope and
replace it with a compound body. A famous early example was
Jones's Improved microscope, which appeared in the 1790s
and owed much to Cuff's earlier aquatic instrument. The
portable Cary microscope which became popular in the 1820s
shows a compound body screwed into a conventional bar-limb
stand of the kind Bancks was manufacturing. The evolution of
the types enables us to put the Bancks instrument into a clear
historical perspective.

The extent of the influence from that distant era on our
own remains considerable. To commemorate the 150th
anniversary of the first identification of the nucleus, I presented a lecture-demonstration on Brown's discoveries at the

Linnean Society in November 1981. The lecture room in Burlington House was crowded with people who had come to see the old microscopes put through their paces, and among the questions that they asked were several relating to the personalities who laid the foundations of modern biology. Around the walls of the hall hang portraits of great figures – Darwin, Brown, the Hookers, Bentham – and I talked a little about them, one by one; and it suddenly became clear just how important those long-lost workers still were. They changed in one's consciousness from being historical figures with a story to tell, into current influences with ramifications everywhere. For example, when Brown died he was still a Linnean Society Vice-President, and his passing forced the Society to adjourn its meeting scheduled for 17 June 1858. The new date was set down for 1 July. By chance, on 18 June Charles Darwin received his momentous letter from Alfred Russel Wallace in which the theory of evolution by 'survival of the fittest' was advanced. Darwin's own work had already been many years in

FIG. 30 The replacement of the single lens cup with a compound body gave rise to the Cary microscope, of which this is a late example (dating from the early 1830s). Apart from the replacement of a wooden box by a cylindrical base, it is a clear descendant of the botanical microscope used by Brown.

preparation, and the arrival of Wallace's letter from Malaya impelled him to act quickly. Darwin realised he could not very well anticipate Wallace's letter by hastily rushing his own views forward, for Wallace would rightly feel that his letter had acted as the stimulus and Darwin's act would have deprived him of priority. On the other side of the coin was the converse argument, namely, that Darwin's work was already so far advanced that Wallace's brief account did not warrant priority in that manner. The re-scheduled Linnean Society meeting provided the opportunity for a compromise solution: Darwin

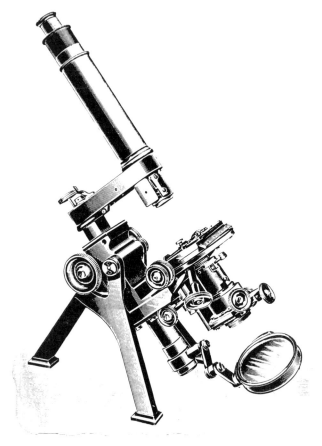

FIG. 31 The famous Powell & Lealand No. 1 microscope of the 1870s shows the compound microscope at its zenith. There is now a tripod foot, but the body-tube still screws into a bar-limb exactly like the simple microscopes we have seen in Figs 24, 28 and 30. The mechanical and optical complexity of these instruments finally made the microscope respectable.

rushed through an outline of his own work, quoting his existing notes and correspondence, and Hooker and Sir Charles Lyell pressed him to present this summary together with Wallace's letter at the meeting of 1 July. His words were written in a more compulsive and attractive style than he planned to use for his book on the subject, and so the concept of the origin of species by means of natural selection received an airing sooner, and in a more popular form, than would otherwise have been the case. Among those convinced by Darwin and Wallace's theory was Thomas Henry Huxley, grandfather of Sir Julian Huxley and great-grandfather of today's President of the Royal Society Sir Andrew Huxley.

And the 'line of evolution' that the microscopes established? That remains with us to the present time. Today's most advanced research microscopes have the same square-shaped frame of the bar-limb construction. So it will no longer do to regard those microscopes by Bancks and his contemporaries as minority-interest dissecting instruments of passing relevance and with an ephemeral history. Like the people who used them in advancing the front of scientific understanding, these little devices have left their mark on today's microscope design. Everyone should understand the essential truth that Leeuwenhoek, and those who followed him, did not make his great discoveries in spite of his simple microscopes, but *because* of them. And when we look at those little-known 'aquatic' microscopes from one and a half centuries ago we should no longer see out-dated little relics from a quaint era that few can now imagine: the design established then is a mainstay of modern microscopy.

Yes, the microscope may have been simple – but the consequences were complex, and perhaps more far-reaching than any of us, at first glance, might have guessed.

Selected References

Anonymous, Leeuwenhoek, who discovered our greatest enemy, *Educational Focus*, Bausch & Lomb, New York, March 1932. [Gives name as *Antoni*, with *Antonj* as alternative.]

Arber, A., Tercentenary of Nehemiah Grew, *Nature*, 147: 630, 24 May 1941.

Baker, H., An account of Mr Leeuwenhoek's microscopes, *Philosophical Transactions of the Royal Society*, XLI: 503 (1740). This date may be erroneous; other evidence suggests this paper was published around 1744.

Baker, H., *The Microscope Made Easy*, Dodsley, London (2nd edn), 1743.

Baker, J. R., The discovery of the uses of colouring agents in biological microtechnique, *Monographs of the Quekett Microscopical Club*, Williams & Norgate, London, 22pp, 1945. (Refers to L's discovery of crocus solution, doubtless saffron, to increase visibility of nerve structures.)

Beck, C., *The Microscope*, R. & J. Beck & Co., London, 1921.

Beltran, F., Dujardin, Felix., etc., *Revista de la Sociedad Mexicana de Historia Natural*, II (2&3): 221–32, September 1941.

Beydals, Petra, Leeuwenhoek-brief No. 27, *Ned. Tijdschrift voor Geneeskunde*, 77 (5): 522–7, 4 February 1933.

Beydals, Petra, Twee Testamenten van Antony van Leeuwenhoek, *Ned. Tijdschrift voor Geneeskunde*, 77 (9): 1021–33, 4 March 1933.

Boltjes, T. Y. Kingma and Gorter, C. J., On the influence of the Condenser, *Nederland Akademie van Wetenschappen*, XLV (8): 3–8, 1942.

Boltjes, T. Y. Kingma and Gorter, C. J., Some experiments with Blown Glasses, *Antonie van Leeuwenhoek*, 7: 61–76, 1941. (Describes experiments with melted beads, see Zuylen, J. van., *Journal of Microscopy*, 121 (3): 309–28, 1981.)

Bouricius, L. G. N., Anthony van Leeuwenhoek, de Delftsche natuuronderzoeker (1632–1723), *De Fabrieksbode, Opgericht door J. C. van Marken*, 44 (10): 3pp, 7 March 1925.

Bradbury, S., *The Microscope Past and Present*, Pergamon, Oxford, 1968.

Brown, R., papers published in *Transactions of the Linnean Society of London*; also privately printed by the Author; reprinted by the Ray Society.

Burton, W., *An Account of the Life and Writings of Herman Boerhaave*, Linton, London (2nd edn), 1746.

Carbone, D., *Contributo alla Storia della Microbiologia*, 38pp, Milan, 1930.

Carpenter, W. B., *The Microscope and Its Revelations*, Churchill, London (3rd edn), 1862.

Chapman, A. C., The Yeast Cell, what did Leeuwenhoek See? *Journal of the Institute of Brewing*, NS XXVIII (433): 5 August 1931. (NOTE: A paper with the same title appeared as: Svihla, George, *The Microscope*, vx: 289–300, 1967.)

Cittert, P. H. van, The van Leeuwenhoek microscope in possession of the University of Utrecht, *Proceedings, Koninklijke Akademie van Wetenschappen te Amsterdam*, XXXV (8): 1062–3 + plates 1 & 2, 1932, and part II, XXXVI (2): 194–6 + plates 1, 3, 4, 5 (2 in text), 1933.

Cittert, P. H. van, *Descriptive Catalogue of the . . . microscopes . . . in Utrecht University Museum, etc.*, Noordhoff N. V., Groningen, Holland, 110pp, 1934.

Clay, R. S. and Court, T. H., *The History of the Microscope*, Charles Griffin & Co., London, 1932.

Cohen, Barnett, On Leeuwenhoek's Method of seeing Bacteria, *Journal of Bacteriology*, 34 (3): 343–6, September 1937.

Cole, F. J., Jan Swammerdam, *Nature*, 1–39: 218, 6 February, and 139: 287, 8pp, unpaged, February 1937.

Cole, F. J., Leeuwenhoek's Zoological Researches, *Annals of Science*, 2 (1): 1–46 and 2 (2): 185–235, with note on Swammerdam's home and plate, 1937.

Cole, F. J., Microscopic science in Holland in the 17th Century, *Quekett Microscopical Club Journal*, [IV] 1 (2): 59–78, October 1938.

Cosslett, V. E., *Modern Microscopy*, Bell & Sons, London, 1966.

Davis, T., *Preparation and Mounting of Microscopic Objects* (2nd edn), Allen & Co., London, 1873.

Digges, T., *Pantometria*, H. Bynneman, London, 1571.

Disney, A. N., *et al.*, *Origin and Development of the Microscope*, Royal Microscopical Society, 1928.

Dobell, C., *Antony van Leeuwenhoek and his 'Little Animals'*, Bale & Danielsson, London, 1932.

Elias, B. [ed?], *Tentoonstelling Anthony van Leeuwenhoek*, Museum voor het Onderwijs te s'Gravenhage, 42pp, November 1932–January 1933.

Faucci, U., *Contributo alla storia della Dottrina parassitaria delle Infezioni . . .* 60pp, Siena, 1936.

Folkes, M., Some account of Mr Leeuwenhoek's curious microscopes, *Philosophical Transactions of the Royal Society*, XXXII: 446, 1724.

Ford, B. J., *The Revealing Lens: Mankind and the Microscope*, Harrap, London, 1973. *The Optical Microscope Manual – Past and Present Uses and Techniques*, David & Charles, UK; Crane, Russak, New York; Reed International, Australia, 1973.

Fred, E. B., Antony van Leeuwenhoek – on the three hundredth anniversary of his birth, *Journal of Bacteriology*, XXV (1): 1–18, 1933. (An account doubting whether Leeuwenhoek could have seen as much as claimed.)

Goch, H. A. van, *Antony van Leeuwenhoek*, undated, privately printed, 18pp, ?1930s.

Goldsmith, O., *A History of the Earth and Animated Nature*, Blackie & Son, Glasgow, and London, ed. A. Whitelaw, 1840.

Gray, A., Memorial of George Bentham, *American Journal of Science*, XXIX: 103–18, 1885.

Hertwig, O. (trans. Mark, E. L.), *Textbook of Embryology of Man and Mammals*, Allen, London, 1912.

Hoeppli, R., Curiosities in Human Parasitology, *Chinese Medical Journal*, XLVII: 1200–13, 1933.

Hooke, R., *Micrographia*, Martyn & Allestry, London, 1st edn 1665; 2nd edn, 1667; reprinted, Dover Publications, New York, 1961.

Hoole, S., *The Select Works of Antony van Leeuwenhoek*, 2 vols, London, 1798 and 1807.

Jackson, B. D., *George Bentham*, Dent, London and Dutton, New York, 1906.

Kluyver, Prof., Herdenking Antony van Leeuwenhoek, *Verslag van de Vergardering der Nederlandsche Vereeniging voor Microbiologie, Overdrukt uit het* Nederlandsche Tijdschrift voor Hygiëne, Microbiologie en Serologie, Doesburgh, Leiden, 1933.

de Kruif, P., *Microbe Hunters*, Cape, London & Toronto (2nd edn), 1930.

Leeuwenhoek, A. van, *Collected Letters*, Swets & Zeitlinger, Amsterdam, 1939–present.

Lewis, F(rederick) T(homas), The Introduction of Biological Stains [by Vieussens & Leeuwenhoek] *Anatomical Record*, 83 (2): 229–63, June 1942.

Lorentz, H. A., Keesom, W. H., de Haas, W. J., Crommelin, C. A., van der Hoeve, J., with Seters, W. H. van, Anthony van Leeuwenhoek [introduction to film made by Seters, q.v. pp. 48–52], *Lectures on Physics and Physiology for American Students*, Leiden; Sijthoff Co., July 1926.

Martin, J. H., *Microscopic Objects Figured and Described*, van Voorst, London, 1970.

Meyer, A. W., Leeuwenhoek as an experimental biologist, *Osiris*, III (1): 103–22, 1937.

Morre, G., *Description of the Principal Tombs in the Old Church at Delft*, 5th edn, 15pp, Koumans, 1912.

Nijnanten, A. van., Schierbeek, A., Leersum, E. C. van, Crommelin, C. A., Boeke, J., Seters, W. H. van, papers for 'Leeuwenhoek Number', *Natura*, 10 (409), 1932.

Parker, G. H., Anthony van Leeuwenhoek and his microscopes, *Science Monthly*, XXXVII: 434–41, 1933.

Poot, H. K., Antony van Leeuwenhoek, *De Meidoorn*, 4 (10): 145–60, October 1932.

Prescott, F., Spallanzani on Spontaneous Generation and Digestion, *Proceedings of the Royal Society of Medicine*, XXIII (4): 495–510, February 1930.

Punnett, R. C., Ovists and Animalculists, *American Naturalist*, LXII (November–December): 481–507, 1928.

Richardson, B. W., Antony van Leeuwenhoek and the origins of histology, *The Asclepiad*, Longmans Green, II (VIII) 319–46, October 1885, reprinted 1900, in modified form.

Rijnberk, G. van, Een verloren brief van Leeuwenhoek teruggevonden, *Ned. Tijd. Genees.*, 77 (5): 478–9, 4 February 1933.

Rijnberk, G. van, De Taal van van Leeuwenhoek, *Ned. Tijd. Genees.*, 77 (22): 2508–10, 3 June 1933.

Rijnberk, G. van and Rijnberk, M. van, Leeuwenhoeken-brieven, *Ned. Tijd. Genees.*, 78 (48): 5463–70, 1 December 1934.

Rooseboom, Maria, Bijdrage tot de Kennis der Optische Eigenschappen van Eenige Microscopen van val Leeuwenhoek, *Ned. Tijd. Genees.*, 83 (1): 63–70, 7 January 1939.

Rooseboom, M., *Microscopium*, Museum for History of Science, Leiden, Netherlands, 1951.

Rooyen, A. J. Servaas van, Antoni van Leeuwenhoek door Leuven's Hoogeschool Gehuldigd [in] *Album der Natuur*, 12: 380–4, September 1904.

Schierbeek, A., Een paar Nieuwe Bijzonderheden over van Leeuwenhoek, *Ned. Tijd. Genees.*, 74 (31): 3891–9, 2 August 1930.

Schierbeek, A., Leven en werken van Antony van Leeuwenhoek, *Ned. Tijd. Genees.*, 76 (45): 5149–63, 5 November 1932.

Schierbeek, A., Uit de eerste eeuwen der mikroskopie, *Vragen van den Dag*, XLIX, 505–28, August 1934.

Schierbeek, A., Jan Swammerdam, *Ned. Tijd. Genees.*, 10: 1025–50, March 1937.

Seters, W. H. van, Antony van Leeuwenhoek, *Vakblad voor Biologen*, 7: 117–23, March 1933.

Seters, W. H. van, Leeuwenhoek-ceramiek, *Ned. Tijd. Genees.*, 79 (14): 1583–7, 6 April 1935.

South, J., Correspondence re the missing microscopes, Royal Society File MC5. See also *Minutes of Council*, 2: 319, 360, 1846–58; 3: 162, 1858–69.

Turner, G. L'E, *Essays on the History of the Microscope*, Senecio, Oxford, 1980.

Williams-Ellis, Annabel [Mrs Clough Williams-Ellis], *Great Discoverers*, BBC Transcript, 32pp, Leeuwenhoek features pp. 13–15, 1929.

Willnau, Carl, Ledermüller und v. Gleichen-Russworm, *Kurt, Scholtze Nachf.*, 24pp, 1926.

Withering, W., *An Arrangement of British Plants*, 4 vols, R. Scholey, London (6th edn), 1818.

Woodruff, L. L., Baker on the microscope and the polype, *Science Monthly*, 213–26, September 1918.

Woodruff, L. L., Hooke's Micrographia, *American Naturalist*, LIII: 247–64, May–June [1919] (printed as 1619).

Woodruff, L. L., History of Biology, *Science Monthly*, 253–81, March 1921.

Woodruff, L. L., Louis Joblot and the Protozoa, *Scientific Monthly*, XLIV: 41–7, 1937.

Woodruff, L. L., Microscopy before the Nineteenth Century, *American Naturalist*, LXXIII: 485–516, November–December 1939.

Wright, Sir A. E., *Principles of Microscopy*, Constable & Co., London, 1906.

Wythes, J. E., *The Microscopist: A Complete Manual*, Lindsay & Blakiston, Philadelphia, 1851.

Zuylen, J. van, The Microscopes of Antoni van Leeuwenhoek, *Journal of Microscopy*, 121 (3): 309–28, 1981.

Published Research – Interim Bibliography

Veitch, Andrew, Find shows evidence of cell studies in 1600s, *Guardian*, 28 July 1981.

Day, Sir Robin, interview with B. J. Ford on Leeuwenhoek specimens. BBC, London, 1.15pm, 29 July 1981.

Ford, Brian J., Leeuwenhoek's specimens discovered after 307 years, *Nature* 292: 407, 30 July 1981.

Ford, Brian J., Found – van Leeuwenhoek's original specimens, *New Scientist*, 91: 301, 30 July 1981.

Berry, Adrian, Find sheds light on microscopy, *Daily Telegraph*, p.5, 30 July 1981.

News report, Historic find in vaults, *Scotsman*, 30 July 1981.

Ford, Brian J., Robert Brown and the nucleus, *Newsletter of the Linnean Society*, 2–3, summer 1981.

Walgate, Robert, Origineles coupes van Leeuwenhoek verrassend dun, *NRC Handelsblad*, 12 August 1981.

Ford, Brian J., The van Leeuwenhoek specimens, *Notes and Records of the Royal Society*, 36 (1): 37–59, August 1981.

Ford, Brian J., Presentation and exhibit, Leeuwenhoek specimens, Natural History Museum, 25 August 1981 [reported in : *Quekett Microscopical Club Newsletter*, 21: la, March 1982].

Ford, Brian J., Correlated optical and electron micrographs of a section by van Leeuwenhoek [highly commended in] Biennial Micrographic Competition, Royal Microscopical Society, 15 September 1981.

Report on discovery of the specimens [in] *Science Digest*, September 1981.

Ford, Brian J., [review of] Essays on the history of the microscope, by Gerard Turner, *The Microscope*, 29: 108–9, 1981.

Ford, Brian J., Specimens from the dawn of microscopy, *The Biologist*, 28 (4): 180–1, 1981.

Ford, Brian J., A clear case of second sight, *Guardian*, p.22, 22 October 1981.

Temple, Robert K. G., Tiny Tomb [in] Science People, *Discover*, 2 (10): 88, October 1981.

Knox, Peter, Ancient Lens, *Observer* Colour Magazine, p.27, 22 November 1981.

Ford, Brian J., Ce qu'observaient les premiers microscopistes, *La Recherche*, 12 (126): 1147–9, 1981.

Ford, Brian J., The annotation of Leeuwenhoek's packets of specimens, a preliminary account, *Proceedings of the Royal Microscopical Society*, 16 (6): 393–5, 1981.

Ford, Brian J., Lo que observaban los primeros microscopistas [trans. of paper in *La Recherche*, q.v.] *Mundo Científico*, 1 (9): 1037–9, 1981.

Ford, Brian J., The mystery of the microscopes, *Illustrated London News*, 269: 76–7, December 1981.

Rogers, James T., What Leeuwenhoek saw, *Scientific American*, 246 (1): 79–80, January 1982.

Cutler, D. F., and Kermack, D. M., Proceedings of Special General Meeting of 5 November 1981, *Newsletter of the Linnean Society*, 21 January 1982.

Ford, Brian J., Bacteria and cells of human origin on van Leeuwenhoek's sections of 1674, *Transactions of the American Microscopical Society*, 101 (1): 1–9 [leading paper] 1982.

Ford, Brian J., Television interview with Phillipe Deguent, Paris, 15 March 1982.

Ford, Brian J., The lost treasure of Anton van Leeuwenhoek, *Science Digest*, 90 (3): 88–92 and 110, March 1982.

Ford, Brian J., View of Leeuwenhoek's original sections through the Utrecht microscope [third prizewinner in] Annual International Photographic Competition, Institute of Science Technology, April 1982.

Ford, Brian J., Cytological examination of Leeuwenhoek's first microbial specimens, *Tissue and Cell*, 14 (2): 207–17 [invited contribution and lead paper], 1982.

Ford, Brian J., The specimens of Antony van Leeuwenhoek. Rediscovering early microscopy [first formal lecture on the specimens], at Facultad de ciencias, Universitad Autónoma de Barcelona, 29 May 1982.

Report, A vision rediscovered, *Science Now* magazine, 2 (20): 556–9, 1982.

Ford, Brian J., The nucleus and the simple microscope, *Journal of Biological Education*, 16 (4): 281–5, autumn 1982.

Announcement, Leeuwenhoek Commemoration Lecture, *Biologist*, 29 (4): September 1982.

Announcement, The world's oldest miscroscope specimens, *The Times*, Information Service, Today's Events, 25 September 1982.

The world's oldest microscope specimens [special lecture], British Museum (Natural History), poster, 25 September 1982.

Knight, Sam., A mystery is solved [lecture report], *Western Mail*, 27 September 1982.

Ford, Brian J., The origins of plant anatomy – Leeuwenhoek's cork sections examined, *International Association of Wood Anatomists' Bulletin* [invited contribution], 3 (1): 7–10, 1982.

Ford, Brian J., Correlated optical and electron microscopy of Leeuwenhoek's elder-pith sections [short technical note], *Journal of Microscopy*, 128 (2): 211–13, 1982.

Ford, Brian J., The rotifera of Antony van Leeuwenhoek, *Microscopy*, 34 (5): 362–73, 1982.

Ford, Brian J., Revelation and the single lens, *British Medical Journal*, 285: 1822–4, December 1982, reprinted as pp.1–7.

Ford, Brian J., Antony van Leeuwenhoek's sections of bovine optic nerve, *The Microscope*, 30: 171–81, 1982.

Ford, Brian J., [Historic report] Leeuwenhoek's original specimens examined after 307 years (revised version of paper in *Nature*), *Olympus Scientific Review*, 3 (1): 10–12, 1982.

Ford, Brian J., What were the missing Leeuwenhoek microscopes really like? *Proceedings of the Royal Microscopical Society*, 18 (2): 118–24, 1983.

[in] Record of Proceedings, *Biological Journal of the Linnean Society of London*, 18 (4): 404, December 1982.

Ford, Brian J., The restoration of Robert Brown's first botanical microscope, [leading paper] *Microscopy*, 34 (6): 406–18, February 1983.

Ford, Brian J., Bacteria and cells of human origin on Leeuwenhoek's sections of 1674 [revised report of *Transactions* paper, q.v.], leading paper, *Bioscience*, 34(2): 106, 1983.

[in] Nester, E. W., Roberts, C. E., Lidstrom, M. E., Pearsall, N. N. and Nester, M., *Microbiology* (3rd edn), pp.2–9, plates 1–2 a and b, 1–6, (note also citation of *Microbiology and Food*, p.776), Sanders College Publishing, Philadelphia and New York, 1983.

[in] Mann, R. D., *Modern Drug Use – an enquiry based on historical principles*, MTP Press, 1984.

Ford, Brian J., New light on the origins of biological microscopy, a presentation at the Royal Society, London, 10 May 1984.

Ford, Brian J., Horace Dall, Optical Craftsman, *Proceedings of the Royal Microscopical Society*, 19 (2): 98–100, 1984.

[in] de Duve, C., *A Guided Tour of the Living Cell*, Scientific American Books, 1984.

[in] Darnell, J. E. jr, *Molecular Cell Biology*, Scientific American Books (in press).

Ford, Brian J., Leeuwenhoek revisited [address to] Inter Micro 84, McCormick Center, Chicago, 1984.

Ford, Brian J., Una ojeada electrónica a los animálculos de Leeuwenhoek, *Rassegna* (Central American Medical Journal), 2: 12–15, 1984.

Ford, Brian J., The microscopes of Robert Brown (historical review), *The Linnean* (in press).

Ford, Brian J., Dioptric miracles, special public lecture, British Museum (Natural History), 10 November 1984.

Ford, Brian J., Leeuwenhoek specimens, a retrospective, student lecture, University College Cardiff, 27 November 1984.

Name Index

Subject Index